THE PLANK ROAD
AND
THE FIRST RAILROAD

FROM MARQUETTE, MICHIGAN
TO
NEGAUNEE & ISHPEMING

Robert D. Dobson

PUBLISHED BY
DOBSON PUBLICATIONS
224 SHORELINE DRIVE
NEGAUNEE, MICHIGAN 49866
2007

The Plank Road
and
The First Railroad

From Marquette, Michigan
to
Negaunee and Ishpeming

by
Robert D. Dobson

FIRST EDITION
First Printing, May, 2007

ISBN 10 and 13
0-9747708-7-6
978-0-9747708-7-1

Dobson Publications
224 Shoreline Drive
Negaunee, Michigan 49866

Published in the United States of America

Chapters

Robert D. Dobson

Dobson Publications

Local History Books

The Railroad That Never Ran
Growing Up in the Salisbury Location in the 40's and 50's
Ishpeming's Salisbury Mine
The Early History of a Mining Town
NEW: THE PLANK ROAD AND THE FIRST RAILROAD

E-mail: rddobson@vbe.com

WE INVITE YOU TO SEE OUR WEB PAGE SOON

For those who like to do research of the Marquette County area, we offer a CD that has about 500 pages of typed notes of issues of several early newspapers, beginning in 1867, but contains mainly the notes of all existing Ishpeming IRON ORE newspapers from 1879 to 1920. by using EDIT and FIND, the notes can be searched infinitum.

We also have two biographies: Ishpeming's Will Bradley (Illustrator)
and Ishpeming's J. Maurice Finn
(Newspaperman, Gold Investor)

Or write to: Dobson Publications
Robert Dobson
224 Shoreline Drive
Negaunee, MI. 49866

Books are also available at Marquette County Bookstores and Giftshops

Bibliography And Works Cited

1. Armstrong, W. F. Railroad History of Marquette County. Marquette: The Daily Mining Journal, 1.23.1932.
2. Armstrong, W. F. Old Strap Iron Line is Found. Development Bureau News, 1.1.1932.
3. Benison, Saul. Railroads, Land and Iron. 1954.
4. Brotherton, R. A. Discovery of Iron Ore. Lansing: Michigan Historical Magazine, April-June 1944.
5. Cadarette, N. "A Short History of the Duluth South Shore and Atlantic Railway," 1928, 1929.
6. Carter, Jim. "The U.P.'s First Steam Railroad." Marquette, Marquette Monthly, 6.2001
7. Casey R. J., and Douglas, W. A. S. Pioneer Railroad. New York: Whittlesey House, 1948.
8. Crowell, and Murray. The Iron Ores of Lake Superior. Cleveland: The Penton Press Co, 1927.
9. Durocher, Aurele A., "The Railroads of the City of Marquette." The Soo News, July 1984.
10. Geological Survey of Michigan's Upper Peninsula. V. VI, 1869-1873, pp 21-23.
11. Hatcher, Harlan. A Century of Iron and Men. Indianapolis: Bobbs-Merrill Company, Inc., 1950.
12. Hewitt, M. L., Cleveland Iron Mining Co. Deed. Marquette County: Register of Deeds Office, 1.26.1884.
13. Historic Resources of the Iron Range in Marquette County, Michigan, 1844-1941. Marquette: Economic Development Corporation of the County of Marquette, 1991.
14. History of Upper Peninsula of Michigan. Western Historical Co., 1883 (Iron Mountain: Mid Peninsula Library Federation Reprint, 1972)
15. Iron Mountain Rail Road. Lansing: Michigan Pioneer Collections vol. 3, 1879-1880.
16. LaFayette, Kenneth D. Flaming Brands, Fifty Years of Iron Making. Marquette: Northern Michigan University Press, 1977.
17. Marquette: The Lake Superior Journal, 1857.
18. Ontonagon: Lake Superior Miner, 1857.
19. Pioneer Iron Co to Iron Mountain R.R. Co. Marquette: Records of Deed Office, 1857.

20. Pioneer Iron Co. to J. L. Reynolds. Marquette: Register of Deeds, Sept. 1862.
21. Saut Ste. Marie: The Lake Superior Journal, 1854.
22. Saut Ste. Marie: The Lake Superior Journal, 1853.
23. Saut Ste. Marie: The Lake Superior Journal, 1855.
24. Saut Ste. Marie: The Lake Superior Journal, 1856.
25. Saut Ste. Marie: The Lake Superior Journal, 1857.
26. Smith, Fred, and Peterson, Karol. The Plank Road. Marquette: Pride Printing, 1998.
27. State of Michigan. An Act to Incorporate the Iron Bay and Carp River Plank Road Company. Lansing: Legislature Annual Session, 1850.
28. Stone, Frank B. Philo Marshall Everett. Baltimore: Gateway Press, Inc., 1997.
29. Swineford, A. P. "History of Iron Mining on Lake Superior." Marquette: The Mining Journal, 1.27.1869.
30. Westren, John Esq. Jackson Company its History and Resources. Saut Ste. Marie: Lake Superior Journal, 8.14.1850.
31. White, Peter. "The Iron Region of Lake Superior." Collections of the Pioneer Society of the State of Michigan. Vol. 8. Lansing: Wynkoop, Hallenbeck, Crawford Co., 1907.
32. Williams, Ralph D. The Honorable Peter White. Cleveland: The Penton Publishing Co., 1907.

It will save the reader, who wishes to pursue any of the above material, much time to first inquire at the J. M. Longyear Research Library in Marquette. Much of the newspaper material, and some of the other material besides, have been already placed in chronological order in separate loose-leaf binders by Mr. Wesley Perron. In a day before computers and copy machines, Mr. Perron located many articles in newspapers and court records, and typed them up, word for word, on his typewriter. This book includes his research, and I am glad to put his work, as well as my own, into print.

The photograph on the cover is of the Plank Road (The Iron Mountain Railway) just before it turns north west to the Deadman's Curve area.

The Referee's Award "A" is most useable if one plots the Rail Road route of Appendix "B". It tells how many feet the Road is from section corners, and gives the Station Numbers. These are the same station numbers used in Appendix "A". Use a topographical or other map.

INTRODUCTION

Ever since I was a small boy, I had heard about "The Plank Road." I remember an uncle talking to his brothers about how you could still see the wood, and how he had seen it. I don't remember where or how. It was in the late 1940's. I was nine or ten. I am now 67 years old, and in the past three years I have begun both researching it, and walking it up and down, and often rewalking. I waited for spring and fall mostly, when the leaves were off the trees and there was not much snow, to go out and take photos. I had numerous friends accompany me and others also helped me with this project. My thanks to Mike Lempinen, Clifford Stammer, Jim Thomas, Dan Landmark, Leo LaFond, the Dave Dompierres, Sr. and Jr., Paul Gravedoni, Eric Anderson, and Roland Koski.

There have been a lot of articles and a lot of research done by others on both the Plank Road, and the First Railroad. I have already mentioned Mr. Wesley Perron who typed up much first hand material, right from the newspapers of the time. But as far as I know, no one has actually traced them by walking them as far as possible, or has written a book about either the The Iron Mountain Railway (Plank Road) or the Iron Mountain Rail Road of Marquette County.

This book is as thorough as I could get, using the Longyear Research Library in Marquette for first hand accounts of what was happening starting in 1845. The Lake Superior Journal, forerunner of the Mining Journal of Marquette, was already being printed in 1850, but it was then located in Sault Ste. Marie, often spelled Saut. Whenever we use the words "paper" or "newspaper" we are referring to the Lake Superior Journal.

My thanks to Rosemary Michelin, who located materials for me on many occasions, at the Longyear Research Library. My thanks also to the Negaunee Public Library for the use of their Michigan Room and historical materials. My thanks also to Thomas Friggens and Barry James of the Michigan Iron Industry

Museum in Negaunee, and the place where the first Iron Ore Forge in the Upper Peninsula was built.

I also thank my wife, Ethel, for reading and proofing every word that is in this book. I have a great appreciation for her work and generous offer. This is the seventh book she has done. Any misspellings of names are the fault of myself. Thanks again to Steve Schmeck for my bar code.

I would also like to thank the Negaunee Historical Museum for what I have learned there, and what maps I found there and have been able to copy and use in this book. The Museum is a good place to begin a study of local history, with 28 rooms of wonderful items for men, women, and especially children, and a yearly membership fee of only $10.00. Or, one can donate for a single day visit or become a Life Member for $100. per person. There is a Railroad Room.

Some of the property I have walked was privately owned and where it was posted, I have received permission from owners to be on their property. Other property has ATV and snowmobile trails for public use near or on the Plank Road and Rail Road. Some property is open to the public. I would suggest that the readers who wish to see parts of the Roads for themselves take a Plat, other maps, and this book, along with them.

I am always open to history I may have missed, and there is always more to learn. I would especially enjoy hearing from anyone who knows more information of any of the three "Roads" in this book. I keep a folder of items to add when a new printing is done, and you can also be a part of this history.

In closing, note that the first railroad was legally spelled Rail Road and has been used in the text, but for the book title I have used the traditional spelling. And remember, this entire book of local history took place before the American Civil War.

Robert D. Dobson

Chapter 1

The Iron Mountains And the First Wagon Road

The first thing to remember is that you are first going to be reading about a very strange land, filled with names you never heard before. The familiar names have not yet been thought of, and the towns are not named. In fact, there are no towns. No one was really even trying to discover iron ore.

In 1845 there was a copper "rush" going on in the Keweenaw Peninsula of the Upper Peninsula of Michigan. Boat loads of people were going there, all passing through the Saut, the then current name of the present Sault Ste Marie, Michigan. After bypassing the St. Mary's Rapids, they were on their way to the Keweenaw to find copper. As they traveled along they happened to often spend a night in the area of Presque Isle (The present Marquette City Park) on their trip to the Copper Country. From the Native Americans in the area of Marquette, they heard of a mountain of "bright and shiny" rock to the west. In fact there is an indication that many Upper Peninsula Native Americans knew of the shiny rock, and that it was somewhat of a holy area they did not enter.

The first non-natives to visit the Iron Mountains were the surveying group of Douglass Houghton and W. A. Burt in 1844. One of that group, Harvey Mellon, actually found the first piece of "Iron Ore" which explained their compass deviations in the area. (January, 1904, Iron Ore newspaper) Mr. Philo Marshall Everett was the first person to search out this Mountain for a commercial benefit, but it is very interesting to note that the first white men who were to see the Mountain did not include Mr. Everett, but a Mr. Carr and Mr. Rockwell, who were led by Marji Gesick, the Indian Chief. (Mr. Gesick's original spelling was spelled Manjekijik in government records.) Carr and Rockwell were sent by Everett to L'Anse, where Native American Gesick was at the time. They took a trail that went to the Bright and Shiny Mountain from L'Anse. Chief Gesick did not want to go near the shiny rock area and did not come all the way to the Mountain, but only as far as the west end of Teal Lake. In his recollections, Mr Everett noted that the mountain was "one hundred and fifty feet high." (31, p. 148)

The description of the iron ore was, as it was told them originally, "bright and shiny," and these words found their way into a written report. If you look around the Jackson Pits today in Negaunee, you will still see this shiny rock near the observation area. The best place to see it, however, is at the Cleveland Mountain in Ishpeming on the hill above the old Cleveland School on Division Street in the Cleveland Location. It is called Jasper Ridge today. From the road behind the hill, which is Jasper Street, steps are in the rock wall, and a path goes up to the top which is still covered with this shiny, bright iron ore. It is my understanding that not all of this type of ore is rich enough to be mined. Thus, some remains at the Jackson Mountain.

The first "finders" of the Mountain by Teal Lake, in what is now Negaunee, had to record their square mile. To do this they had to go to Copper Harbor's government office. Teal Lake was not on the map, but they guessed at where it was and the land was obtained by the group from Jackson, Michigan, and thus the shiny, bright hill was called the Jackson Mountain, and later became the City of Negaunee. Other groups went further west and filed claims for what they called the Cleveland Mountain as the group was from Cleveland. A third party wanted some mining land, and a man called Mr. John Burt, filed a claim for property just west of the Cleveland Mountain, named "Little Mountain". That became the Lake Superior Mine. The two latter mines became a part of the city of Ishpeming. An old Mine Office is still at that mine area, at the top of the South Pine

Street hill, and it is still called by some, "The Burt Office Hill", and the building itself is often called the "Burt Office." And we are talking about 1850.

Because no one knew where precisely these land claims were at first by the Government men at Copper Harbor, legal problems arose as to who claimed what exact property. To make a claim, one had to erect a building on the property and have someone in charge. This had been done, and the buildings were often just crude houses. Much of the time no one was on the property at all, and the Cleveland structure was burned down once, and another building was taken down and moved to Marquette since it was usable. Exact property locations on government maps were eventually legally solved and groups were allotted property.

At first as the word spread about these shiny mountains, men were not sure what it was that was bright and shiny, any more than the Native Americans. Some thought it might be copper, or gold along with iron ore. Soon word got out that the mountains were iron ore in Michigan's Upper Peninsula, and men purposely came here to make their fortunes. Associated with the Jackson Iron Company were Philo M. Everett, Abram V. Berry, William A. Ernst, S. W. H. Monroe, and F. Farrand. (8) Tower Jackson was the mining agent for the Cleveland Co. Most of the owners of that company stayed in Cleveland, with communications back and forth. And, of course, Mr. Burt represented the Lake Superior Mine.

The first ore removed from the U.P. was a 300-pound load, carried on men's backs, to Lake Superior, and shipped downstate. A fine bar of iron was made. The first actual mining started at the Jackson Mountain in 1846. (31, p. 148) A locating map is inside the front cover.

In 1847, the Jackson Mine group decided to build a Forge near the Jackson Mountain at the Carp River. As ore contained about 20% or more waste rock, no raw ore would be carried by schooners to furnaces down below the "Saut." In the Forge, after heating the ore chunks red-hot into a soft mass, the Forge would hammer much of the rock impurity out and formed a bar, called a "bloom." To do this, the Carp River was dammed up. The required machinery, including bellows, and hammer, was carried with much difficulty on a Wagon Road built in 1848 from Lake Superior to the Forge. This Wagon Road also continued to the Jackson Mine, and on to the Cleveland Mountain. The following old maps show the route of the Wagon Road. On the first map, the Forge location is in the upper right square. Jackson Mountain, which became Negaunee, is at the south end of the trail. A dashed line goes on to the west to Cleveland Mountain (not shown) which became Ishpeming.

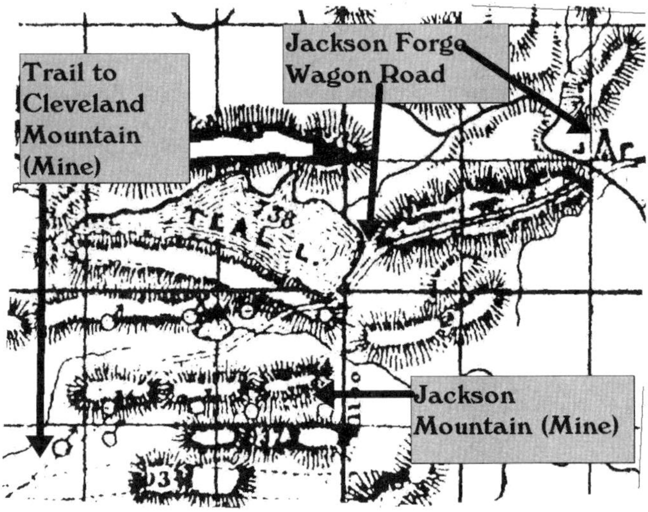

It was difficult to get workers in the early years. Winter was not only cold and housing poor, but once the lake froze over, there was no communication, or food, or necessities obtainable until maybe June. There were other problems at the Forge. The eighteen-foot high dam broke one year and had to be rebuilt. (31, p. 149) A total of twenty-four men and two women wintered in one large cabin in the winter of 1848. (28, p. 73) The managers of the Forge changed often. A Mr. Jones was in charge one year when the men worked for a long period without pay. Mr. Jones left one day for Green Bay, as he was unable to pay them, and never returned.

To get the blooms from the Jackson Forge to a schooner, the Wagon Road built to Lake Superior was used. This is the map of it-----, as continued from the preceding map. Both maps, with other details, can be found in Frank B. Stone's Book: Philo Marshall Everett.

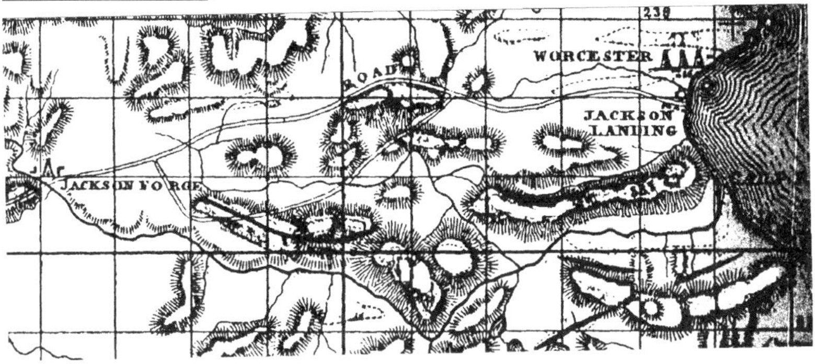

Just a bit east of the Jackson Forge on the above map is a southern road loop that was an improved route of six miles that was made during its use. The "new" six-mile route, or the south loop, went down to Eagle Mills and followed the general County Road 492 route today to just north and east of "Deadman's Curve." The Wagon Road went north of the 'Curve,' rejoined the original road, then curved south and crossed 492 (and the Old Plank Road and Railroad) near the bottom of the S. Curve near Grove St. More about this route will be shown later in this book. Of course, none of these were there when the Wagon Road was built. The Plank and Rail Road really crossed over the Wagon Road.

A Wagon Road map is obtainable from the State of Michign website. We have printed the State map and I have added on the six-mile improved section constructed in 1850. The Plank Road and the first Rail Road took a different route than the Wagon Road, but what is of great importance is that, when the Plank Road and the Iron Mountain Rail Road were built, they crossed the path of the Wagon Road at least five times on their way from the harbor to the Iron Mountains. I believe that one part of the original Wagon Road can still be seen and am putting photos of it in this book.

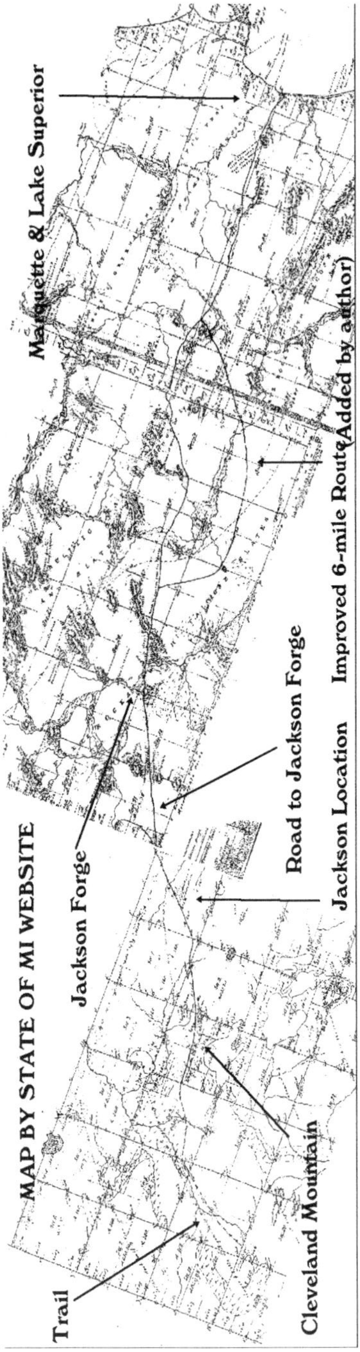

In 1848, the Wagon Road from Lake Superior to the mountains was not working out. From the time the first blooms were made in February of 1848, two six-horse teams were used to haul the iron ore wagons. In the summer the wagon wheels, under loads of over a ton, often sank and dragged through sand for much of the trip. Wagons also had to be used in the winter because sleigh runners, under the heavy loads, went right through the snow and rode on bare ground poorly. (28, p. 75) In addition, all the horses and their feed had to be imported by boat, as there were no farms yet.

There was also the problem of getting through the Saut Rapids (Sault). All the blooms had to be unloaded and carried on a plank road from the top to the bottom of the rapids and then reloaded on to another schooner. It was also found that the loading area by the Carp River (now Marquette) was not the best, as Northeasters made loading impossible. The loading area was sometimes changed to the Dead River area by Presque Isle of Marquette because the Island gave more protection. This river was then called the "Mad River." The new loading area also, however, added three more miles of the poor wagon road. (28, p.74) It is said that the first Forge blooms were sold to a Mr. E. B. Ward and the iron used "to make the walking beam of the steam boat, 'Ocean'." (29)

This area is now being restored by the Michigan Iron Industry Museum on Co. Rd. 492, east of Negaunee. There is a large monument there, built in 1914. See its picture on page 20 from the Iron Ore newspaper, dated September 5, 1914. "Monument Erected by the Jackson Iron Co. to Mark the Spot Where the First Forge, was Built by the Jackson Mining Co. in 1847." The monument is still there and will be accessible to the public when the bridge from the museum across the Carp River is completed in the near future. The old access route is no longer available to the public. It can be seen from the Museum during months when leaves are off the trees. The Wagon Road went to, and from, the Forge going north-east.

This map is from the Michigan History Division (1977.4 C294) A Carp River Forge Report. Note 492 states : Plank Road.

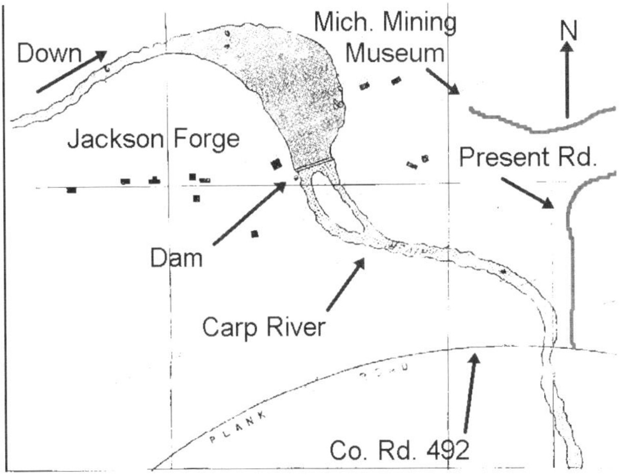

Another new forge went into business, the Marquette Iron Company Forge, in 1849, by Robert Graveraet, and financed by W. A. Fisher, Amos R. Harlow, and Edward Clark, all from Worchester, Massachusetts. The name of the Marquette area was then called Worcester (pronounced Worster) for a time. The name of Marquette for the forge was not named after the city, but after Father Marquette.

The Marquette Forge was purchased by the Cleveland Co. in 1853, but it completely burned in 1855. Therefore, it was gone into history when the Plank Road and the First Railroad went into business. In 1853, the Forest Forge began on the Mad River (Dead River) and down stream, the Collin's Forge also began in 1853. The Collin's was the first forge to become a furnace (in 1858), with a successful experiment to preheat the air to obtain higher temperatures over 3,000 degrees for melting iron ore into a liquid. Many furnaces then came into existence and poured ingots called "pig iron." As slag impurities were skimmed off the liquid, pig iron was of a purer quality than blooms. Just about all of the furnaces eventually went out of business because of a lack of wood for making charcoal. Early in the 1840's and 50's, charcoal was made by enterprising individuals in their farm yards. Later it had to be made in the "rock hives" close to the hardwoods, yet not too far from the furnaces. Here is a most interesting fact: It took 200 tons of charcoal to make one ton of pig iron ingots. Charcoal is made by burning wood without oxygen. When burned later in an iron furnace, the charcoal finally obtained the oxygen it had long sought, and burned very hot. Some furnaces such as the 1863 Morgan and the Greenwood built tram ways to get charcoal from further distances. (16)

According to Peter White in an article he wrote in 1885 about the early years, "there were no white habitation and only about two Indian huts to start with, and a small warehouse of the Jackson Company." Peter White, himself, also arrived at this early time of 1845 and was only 19 years old. He came, he said, "shortly after the first excitement over the discoveries of copper in 1845, to go up to Lake Superior which had gained a great hold on the imagination of the boys of that day." (31, p. 149)

Chapter 2

Thoughts of a Plank Road and a Rail Road

The rough Wagon Road continued into 1850, carrying ore from the Jackson Mine to the Jackson Forge, and on down to Marquette. The Cleveland Mountain also used the road to take their ore down to the Marquette Forge in Marquette for processing into shippable blooms. Shipping continued to be difficult. One year the Jackson Forge had ten tons of blooms ready to go to Marqette, but only two tons got there before the shipping season ended. (28, p. 74) In 1849 John Westren of the Jackson Iron Co. noted that in September there were seven tons of blooms at the Jackson Forge, but only one ton made it during the fall. Also in 1850, he noted that the Forge had spent money to build six new miles of the road to Marquette to avoid a portion of the old road which was "impassable." (29) This was the six-mile Wagon Road addition shown on the previous maps. According to a news article by W. F. Armstrong, the wagons were quite hefty and carried 3,600 lbs or almost two tons. Neither forge made any money, and it was decided that if transportation could be improved, the raw ore itself could be shipped efficiently. (2)

Plank Roads were really very common in America at this time. The road that was being used to transport items around the St. Mary's falls at the "Saut" was a plank road. For plank roads, land was cleared and smoothed off, but there was often little filling of valleys, or digging of channels in higher land. Wood was simply laid side by side and quite a smooth highway was produced, be it up and down in hilly country.

In Marquette, "A plank road corporation was organized early in 1850."(27 p.100) We find that the name of the company was the Iron Bay and Carp River Road Company, and the leaders were Czar Jones of the Carp Forge, Mr. Philo M Everett of the

Jackson Mine, as well as Amos Harlow, and several others. The Plank Road bed was to be 16 feet wide, with the planks 8 feet wide and not less than three inches thick. It was to be a "good, smooth, and permanent road." (27)

In walking all parts of both the Rail Road and Plank Road, they both are always 16 feet wide when measured. Both must have had to follow similar basic rules from the state legislature. They could decide on the grade of wood, and one would think that hardwood would have been used. However, hardwood evidently went for the tons and tons of charcoal needed for the forges and furnaces. We find that when a Mr. Antoine Deloria died in 1909, that his obituary in the October 2nd newpaper stated that he both "worked on the old plank road" and he made ties for the first railroad out of pine cut at Eagle Mills. "He furnished several thousand of the first class ties..." Perhaps planks and ties were both of softwood for the Plank and the Rail Road.

In order for any kind of improved transportation from the Iron Mountains of Ishpeming and Negaunee to work, something would have to be done with St. Mary's Rapids. On August 14th of 1850, the Lake Superior Journal printed: "It is thought an effort will be made in Congress this session to get a canal made at Saut Ste. Marie..." (29)

In 1850, a new person entered the city of Marquette. He was from the Eastern United States and he pushed for a Railroad. His name was Heman Ely, and his brother George also came. (Heman and George are buried in Marquette's Park Cemetery, along with other family members.) Heman had a land elevation survey made from Marquette to the Iron Mountains. The population of Marquette County was now 150 persons. He spoke with the men of the Jackson and Cleveland Mountains regarding their using the Rail Road if he built it, and they agreed. A third person, Mr. Burt of the Lake Superior Mine also was a main part of the Rail Road to be built, so he could start up the Lake Superior Mine at the "Little Mountain." (We shall use the incorporated spelling of Rail Road as it also distinguishes it from other railroads.)

Heman did his survey for his Rail Road in the summer of 1851. There were no legal owners of much of the land yet, and Mr. Ely was surprised at how easy the grade would be to build the Rail Road. But his real problem would be financing. The Cleveland and Jackson Companies waited for a full year for the construction of the Rail Road to begin, and then felt that they must begin a Plank Road in order to ship by the year 1855 when the canal would be completed. There was talk in some later Plank Road articles that the Jackson and Cleveland tried to get out of the Rail Road Contract with Mr. Ely. There were some with financial interests who were opposed to the Rail Road. However, we find, that even before the Rail Road was entirely completed that the Cleveland and Jackson Companies were bringing ore to where the Rail Road was completed at the area of what is now Eagle Mills, and already using the lower part of the Rail Road to get their ore to their Lake Superior docks.

Here is a photo of the Wagon Road. More photos in Part II.

A Monument to the First Forge

"This Monument was erected in October, 1914, by the Jackson Iron Company to mark the spot where the first forge on Lake Superior was built by the Jackson Mining Company in 1847."

In the fall and winter months it can be seen from the deck of the Michigan Iron Industry Museum. The photo is from the Museum Library and can be found in the 1914 Proceedings of the Lake Superior Mining Institute, published by the Presses of the Iron Ore, Ishpeming Michigan. It is also found in the IRON ORE newspaper of September 5, 1914. It is quite a large monument.

Chapter 3

The Legal Battle Between the Railway and the Rail Road

When the Cleveland and Jackson (now the Sharon) Companies began the Plank Road, they chose a new name for "The Iron Bay and Carp River Plank Road." It became known as the "Iron Mountain Railway." Those who subscribed to it included: John Westron, L. Bascom, Wm A. Ernst, A. R. Harlow, F. W. Kirkland and Philo M. Everett and Czar Jones.

The Railway was built quickly, from 1853 to 1855. There is little of it left to be seen, and where it is seen, it is often right next to the Rail Road which can be easily recognized. This caused a great legal battle that was not solved until 1855. Mr. Heman Ely, who did the survey for his Rail Road claimed that he had been invaded by the Railway and the two companies who had agreed to support him. At two places the Railroad was forced to cross the Railway, and at another place there was a rock bluff abutting into a narrow gorge that both ore carriers had to pass through and there was only enough room for one, with the Plank Road already being there. Mr Peter White tells us about the survey for the Rail Road, His view of the grade was harsh: "The grades were heavy---rock cutting very extensive, bottomless swamps to cross, with fills numerous and difficult to make." However, the Lake Superior Journal on May 28 of 1851 reported that Mr. Burt "finds it more favorable than he anticipated." Both can be correct in terms of equipment available for building grades. When one walks it today, it has areas that required no grade at all, and some areas of large cuts, and high grades. It wasn't, however, like crossing a mountain range.

One of the strange things that I found when I first started walking the old remains of both the Railway and the Rail Road was that the Rail Road built-up grades and cuts were easily visible, but that the Plank Road path was often invisible. Of course, County Highway 492 is built right on top of some of it. The best visible section of Plank Road is just west of the "S" curve.

Looking at the old maps, the Rail Road and Railway (Plank Road) seemed to be quite a distance apart from each other. They form the double line between Marquette and Negaunee.

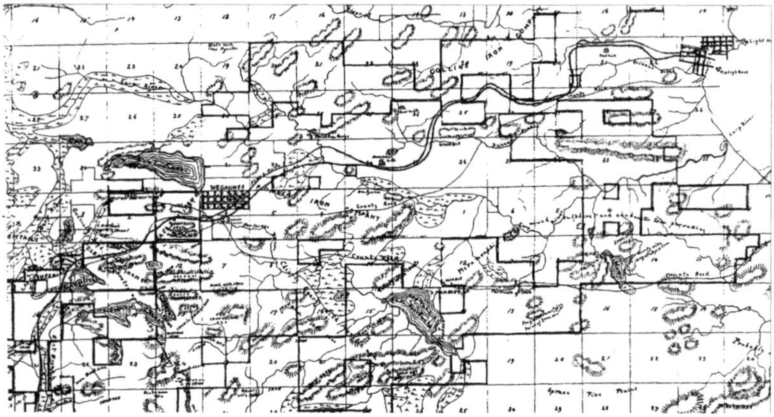

Several times in the beginning of my exploration, my brother-in-law and I, walked into the woods for some distance on both sides of the Rail Road grade to find the Plank Road. We never found it because we did not suspect that the road we were on right next to the visible Rail Road grade was the original Plank Road. It was sometimes still in use as a road by 4-wheelers, snowmobiles, and even trucks and bicyclists, after 154 years, or from 1853 to present 2007. What fooled me the most is the fact that at some points there was only the Rail Road grade, and no Plank Road could be seen, which also threw me off in my thinking that it must be some distance away. Of course, the Plank Road often did not build any grade or cuts, and was just there next to the Rail Road grade, but at ground level, with trees on it, and not presently used by traffic. It was only in the course of my research that I came across the "Referee's Award" that told me the the Plank Road and the Rail Road were often very, very close to each other.

Now that we see how close the two construction projects were, we can understand the legal battle. Here are some of the 1855 decisions made for each of the entities at the outcome of the court case. Both had to follow stipulations, and several times the "Referee's Award" will tell of how the Plank Road (Railway) and the Rail Road are not to be on top of each other, but indeed are to be right next to each other. Appendix "A" will have four pages of the total details of the "Referee's Award. The stations are key marking points for every change of direction. It is Section "B" that gives me the stations where the Rail Road crosses every section line. I mark those on a "sectioned" map and then I can read "A" very easily.

On page one of the "Award" it reads that the Rail Road: "From station 283 to Station 286 it is to be to the southward of the line of the Railway....and distant from it not less than twenty feet from center to center of the respective roads." These two stations are just north-west of dead-man's curve, in a flat area, but with some curves. Note the distance is to be 20 feet from the center of a 4' 10" gauge to the center of a 4' 10" gauge. That includes almost 2 feet, 5 inches, times 2 (Half of each 4' 10"), or 5 feet used up, leaving only 15 feet between the rails of one to the rails of the other. (If the Plank Road rails were less than 4' 10" apart, then the amount of space would be even less.)

Again the Referee's report reads as follows regarding the Railroad from points 356 to 396 (The narrow Gorge area): "it shall have the right to be located by its Engineer at any distance not less than eleven feet from the Railway from center to center." (This only allows 6 feet between the rails of the two roads.

We also know that in this Gorge the Railway (Plank Road) was evidently in the only space available for passage, and that the "single space" was now awarded to the Rail Road. The Railway, at station 388, was required to change its route by cutting and moving "northward a distance of 27 feet from the present center of Railway to outside of cutting." At station 391, "the rock to be cut northward a distance of ten feet from the present center of

Railway." And the Railway was instructed to move northward as far as possible for the Rail Road.

On page two of the Award, we do find how close they are: "It is also a rule that the space between these two roads where such space is not more than twenty feet from center to center shall be held in common by both parties...." It also notes that where the Rail Road is higher or lower than the Railway (Plank Road), that a foot and a half of distance be allowed for every foot of difference in elevation.

Of course, I could not believe how close both of the Roads were when I began to search. If I had seen the Referee's Award from the first, it would have saved considerable time going off in the woods on either side of the Rail Road grade, looking for the Plank Road grade when many times, I was standing on it.

We shall look more at these Referee's rulings when we trace the route of both the Railway and the Rail Road in Section Two. And we shall see where the gorge is, and many more interesting items along the way.

Iron Mountain Rail Road looking north. 492 at Grove Street to right.

Chapter 4

THE PREPARATION AND THE BEGINNING OF THE PLANK ROAD

We return to 1851. There is knowledge that the Sault Lock will be completed in 1855. Excitement is growing. The Lake Superior Journal is telling the world of what is happening, and it goes into other papers as editors read the Journal. In December of 1851 we read (as taken from the Detroit Tribune):

"Iron enough to build Whitney's Pacific Rail Road might be taken from the Mountain and not be missed.... It exists there in such abundance and is of such an extraordinary quality that in a late report of the United States Geologists, this prophecy was made in regard to it. Says the report: 'This region possesses an inexhaustible supply of iron ore, of the best quality, removed from 12 to 30 miles from the lake shore, with a soil by no means sterile, and covered with a heavy growth of maple, yellow birch, pine, and oak—and it is to this source that the Great West will ultimately look for the finer varieties of bar iron and steel.'"

The Sharon Co, which in 1853, now owns Jackson Mountain and Forge, figured up how to make a profit as follows:

Quarrying or picking up loose blocks, per ton: .25
Converting into blooms: $12.00
Transportation to Lake Shore: $1.00
Transportation Lake Shore to Erie, Pa.: $2.00
Transportation Erie to Sharon, Pa.: $1.00
TOTAL COST OF BLOOMS PER TON: $16.25

Mr. Ely was building his Rail Road, but just a bit. He built a house near Whetstone Brook, and an office near by. When questioned about the slow progress by the Sharon Co., he noted to them that he can only go as fast as he "can pay up." It was, for him, a "pay as you go project." It is noted that he "kept steadily digging away." Several meetings followed and he asked for the Cleveland and Sharon Co's to put in funds. They said they could, but wanted full control. He said he only would give "half a control." Talks ended, and they began the Plank Road. (28) There was no cost listed for the building of the Plank Road, but they proposed to build it in the spring of 1852. (It ended up costing $120,000.) (9)

One reason I believe that the Rail Road got such a slow start is that the part from the Lake up to Grove Street was the worst part of the survey upon which to build a grade of only a few percent. The route followed under or close to County Road 492. It was obviously rough for the Plank Road as well, as the following detailed map shows:

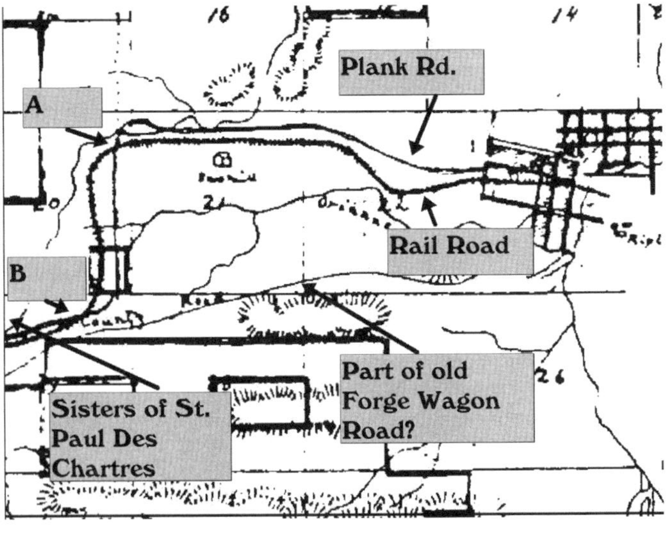

The Plank Road is the thinner line (see arrows), and of course when it was built, there was no Rail Road (the thicker line) this far yet. One can see the ruggedness by the hump found at the top left area, and of course, unlike the Rail Road, the Plank Road tried to just follow the lay of the land as best they could find it, but leveling off hillside areas. It was the Rail Road that had to then battle with the route that was in a sense, stolen from them, and had to cross the Plank Road twice on the "S" curve which still exists on the present 492 (See "A" and "B" on the map). I don't imagine they made any overpasses, and may have just torn the Plank Road out at those two spots. There is no information in the Referee's Award. If the Plank Road was destroyed at the two crossings by the Rail Road, that may be the real reason the Cleveland mine began loading ore on to the Rail Road at the area that eventually became Eagle Mills. The "County Road" shown on the above map is not the Wagon Road from the Jackson Forge, but it does seem unusual that the Wagon Road is not shown on this map, crossing at the bottom of the S curve. It is shown on the next map.

The 1848 wagon route preceded Mr. Ely's survey of 1851 and used Mr. William A. Burt's survey of 1844. The map below shows that the Plank, going west, crossed the Wagon Road at the south end of the big "S" curve by Grove St. in Marquette, crossed again NW of dead man's curve at the top of the Plank Road route, and again just south of the gorge in the large swamp area (left side of the next map).

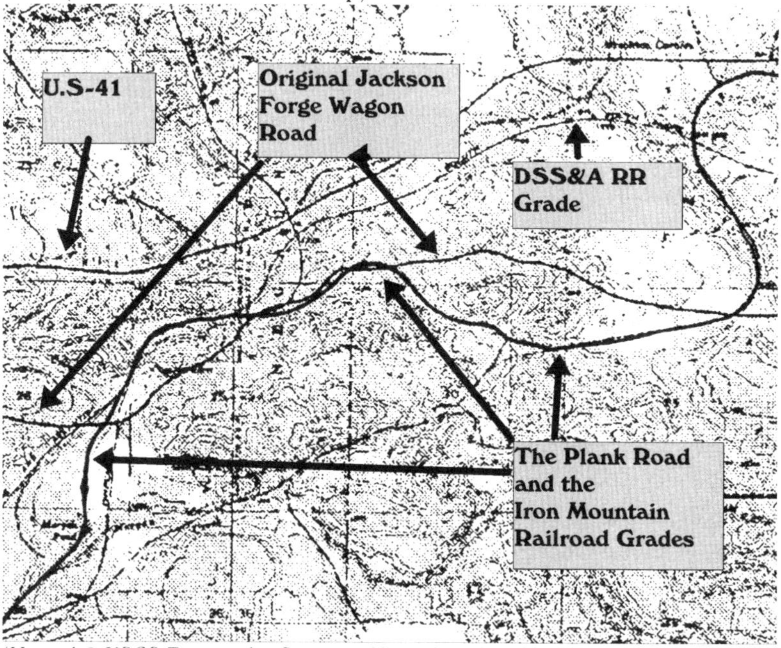

(Maptech® USGS Topographic Series, ©Maptech® , Inc. 603-433-8500, Forge Wagon Road added.

The Sharon Mine and the Cleveland Mine both used the wagon route until 1856, when the Plank Road was completed.

We can ask ourselves, "Why was the Plank Road changed to a Railway? My own guess is that in building it, the wagons on their own did not seem to keep on the eight-foot planks very well. Whether the planks bent under the weight of the wagon wheels allowing them to slip off the edges, or slippery ice or snow caused the same, metal covered wood rails were soon installed to correct one or more of the problems. Remember, the planks are only three inches thick. I don't think they would be very stable under a lot of weight. Plus, when a wheel or two came off the planks, they probably sank into the sand which covers much of the route.

Thus, they tried to improve the Plank Road by the use of wood rails, covered with metal straps. How were they nailed? Some seem to have guessed the straps were nailed down separately, end to end. (4) An advertisement in the newspaper in 1865, tells us that the rails evidently were removed and stored for several years after the Rail Road went into business. The note seems to indicate that they were flat cast iron strips, and nailed end to end on the top of the wood rails, with one nail holding the ends of two strips:

"STRAP RAILROAD IRON FOR SALE: About 12 tons, two and half by five eighths inches, spike holds punched and countersunk, ends of bars notched and matched." (Lake Superior Journal, October 6, 1865.) The ad in its entirety will be shown near the end of this history section.

I have not found the answer of why it was changed to a railway from a Plank Road, except it happened quickly. An 1871 book tells us, "In 1853 the Jackson and Cleveland companies joined in construction of a plank road from the settlement of Carp River to their mines, completing it in 1856, in which year it was converted into a tram or wooden railway." And, with wood rails, how did two pass each other, going and coming?

One would think that the wheels would be ordinary and ride between the planks, with the planks steering them, one on a side of each wheel. However, with metal straps on the rails, the wheels obviously rode on the rails, like a Railroad "T" rail. How then were the wheels fashioned? If there was a ridge on the wheels, were the wheels cast iron? Did ridges break off if the cars jumped the track? Rails were cast iron yet, and they broke quite often in the early days and caused many tragic rail road accidents. In any case, the rails did not seem to work, either, with cars rolling even faster down the tracks and jumping the curves and destroying animals.

In later years, we do find some of the names of four workers who were involved with the Plank Road. It is interesting to note that three of the four were French Canadians. Perhaps they found the winters warmer here whereas others from down below found it colder.

A Mr. Antoine Deloria died about October 2, 1909, and his obituary reads: "Antoine Deloria also worked on the old Plank Road over which ore was transported from the Jackson, Cleveland and Lake Superior mines to Marquette before the completion of the Iron Mountain Rail Road (no ore was carried from the Lake Superior Mine until the completion of the Iron Mountain Railroad in late 1858.) His son, by the same name, came with the family, being born in 1831 and who died in 1926. That obituary notes that the son was "a foreman in charge of a squad of teams engaged in hauling ore from the Cleveland Mine at Ishpeming and the Jackson Mine at Negaunee, over the old Plank Road to the docks at Marquette." It also notes that the family was from "a few miles south of Montreal, Canada.
(Negaunee Iron Herald, 1.29.1926)

Another Negaunee man, Mr. Jacob Dompiere, died in November of 1905 and his obituary was also in the Mining Journal. It notes that "He spent his first year or more in the county of Marquette, having come from Montreal, his native place, on May 30, 1852. ....He was one of the first men to drive a spike on the plank railroad over which the first ore transported from the mines to Marquette was hauled by mule teams. Mr. Dompiere worked on the road the greater part of the time it was under construction, later driving teams over it." The Dompierre descendants still live in the Negaunee area.

A Mr. Frank Zoberlein died in July of 1912, and his obituary appeared in the Ishpeming Iron Ore newspaper, noting that he "helped to construct the Plank Road from Ishpeming to Marquette for early ore shipments." It also noted that the Plank Road ran until 1857 when "Ely's Steam Rail Road was completed and could carry 1,200 tons daily." This article also mentioned incorrectly that the early Railway (Plank Road) ore shipments were from the Lake Superior and Jackson Mines. The newspaper, of course, is in error once again, and should have said, the Jackson and Cleveland Mines. It is easy to see that after 50 years a lot of the facts are already lost to history as by the time of this obituary, the Lake Superior was the largest of the three mines.

Men came from Canada and from areas south, and came to be involved with investing in businesses, and as labor for building docks, forges, sawmills, homes, and all else cities need. In Marquette, the Cleveland Company built a large unloading dock for freight, and then soon extended it and added the warehouse so that arriving goods would be sheltered. There is this fine night description of the town in regard to the Marquette Iron Forge: "Its bright fires as the town is approached at night from the Lake are seen at a great distance, getting nearer, the confused roar of steam engines, mingled with the sound of forge hammers is heard" and adds a note about the "the immense glowing masses of iron."

Two other industries are now opening up in the Marquette area according to the newspaper on November 12, 1853. One is the marble quarries of Burt and Ely, noting that "The marble is veined and of all shades, but generally a rich rose color, unlike any other found in the United States." Also noted was that "It admits of the highest polish and is pronounced, by workers in marble, equal to the best Egyptian." The other industry was the "cutting up and manufacture of oil stones for market" by the Carp River Oil Stone Company.

Sawmills are also showing up in many places. Mr. Burt and Mr. Ely have one at the mouth of the Carp River, and the lumber produced the whole season "has been entirely inadequate to the demand." Graveraet has a saw mill two miles north of Marquette on the Mad River. Two miles west of town Mr. Harlow has a saw mill. A Mr. Butler and Mr. McConnell are building a forge on the Mad River above Graveraet's Mill. They are planning to import lots of coal by ship for their forge. However, the other forges are having charcoal made for their use. This means many, many men are busy being employed cutting hardwood trees to make charcoal.

As the Plank Road began construction in 1853, Mr. Ely sent a letter to the Lake Superior newspaper, showing the great possibilities of his Rail Road and the future of iron. The letter was published on November 12, 1853. Part of his letter noted that "the Iron mines here, as soon as facilities are opened for full access to them, will export annually 2,000,000 tons of iron ore. This would be worth, delivered at Marquette Bay for Shipment, $8.00 per ton, which is 16 Million dollars. Carried to market, and manufactured, yielding readily as it does, sixty percent malleable iron, it is worth at the lowest prices sixty million of dollars."

Mr. Ely is making headway on his Rail Road. It is noted in the paper that the grade has been now cleared and graded for five miles from Marquette. This is noted to be also the most difficult part of the Rail Road. According to the paper, the Rail Road ties are in for the first six miles. (Something wrong here obviously.) The article also notes that the Cleveland Plank Road is being built from Cleveland Mine towards the Lake and that the upper portion is already done.

Also noted is the surveying of an improved wagon road south from Marquette to the Bay De Noquet and on the Green Bay. It will be only 57 miles to Lake Michigan, rather than the current 70. Note is made that it will be extensively used in the winter.

The 1853 letter ends with one thing needed yet in Marquette, and that is a breakwater for the harbor. (22, 11.10.1853)

This photo shows the original road from the Forest and Collins Forges up to the Plank Road as seen from Wright Street, looking east. It is the continuation of the present Forestville Road where it comes from the south and meets Wright St. We are just north of that intersection going towards U.S.-41.

Another photo of the old Jackson Forge & Cleveland Mine Wagon Road in use until 1856.

Chapter 5
The Year of 1854

It must have looked like the Plank Road was a super highway, and many people must have begun using it. Many mills and forges seemed to spring up along it, or have roads to it for personal as well as commercial use. I found it mentioned in two places as "smooth." One story tells of a teamster instructed by Mr. Ely to haul materials on the Plank Road to Eagle Mills, but he found many gates put up, and he could not get around them. Mr. Ely, builder of the Rail Road, told the teamster "I want you to take a load of supplies tomorrow morning to my men at the mills, and never mind the gates." The teamster did quite a bit of chopping and completed his delivery. The next day he was arrested, but the court discharged him and that was evidently the end of any gates. (28, p.102) Much of the Plank Road, of course, became today, the present County Road 492.

In February of 1854, the Plank Road was being completed with the three- inch planks being laid side by side, each eight-foot long. The newpaper noted that the Duncan and Parish steam sawmill was producing up to 15,000 board feet of planks per day. It was on section 34, T48N, R26W in the Eagle Mills area.

You have probably never heard of the Eureka Iron Company Mine. I never had either until researching this book. Yet in 1854 it began to appear in the newspaper articles, and continued to do so through 1929. Its actual location seems to move around a bit from article to article.

In the February 1, 1854, issue of the paper, it is noted that the Eureka company "have a water power sawmill nearly completed and quite a good force, chopping cord wood for coaling next summer. Their works are about two miles from the village on the Iron Mountain Road (The Plank Road?). They are raising iron ore at their place, but it is of inferior quality; but it is said to be sufficiently rich for making iron in a blast furnace to a good advantage..." More about this mine later.

In the June 3, 1854, edition of the paper, Peter White wrote a letter to the editor noting that Mr. Heman B. Ely has "not done much since last fall toward completing their Rail Road to the mountains." He does note that it might be ready by the following spring (very incorrect). He notes also that the Cleveland Iron Mining Co., in Ishpeming has shipped down to the Lake already, 12 tons of ore on the Wagon Road, and have 70 tons of blooms at the boat dock, presumably made at the Marquette Forge. The Sharon (Jackson) Iron Co. is busy building the Plank Road, and has by this time of June 3, already graded half of the distance, with Peter White noting that the grading includes "the heaviest and most difficult part." He notes also that the Sharon Co is "building an unusually large wharf and break water at the north end of the harbor.....out into 15 feet of water." The breakwater, according to Mr. White will be extended another hundred feet from the end of the dock.

The Letter to the Editor also notes that the Cleveland Company is anxious for its end of the Plank Road, and is building its end from the Sharon Company to its Mine (Negaunee to Ishpeming), and also is building a separate dock, "seven hundred feet on the south side" of the Sharon dock for its own use. Peter White ends his informative letter with the sentence regarding the two ore docks, that they: "make in all, fifteen hundred feet front, with fifteen to eighteen feet of water – room sufficient at least twelve sail or steam vessels to load or unload in any weather." (21, 6.3.1854)

The newspaper reports on June 24, 1854, that both the Rail Road and the Plank Road were getting built. However, it was at this time the injunction against the Plank Road was obtained "for running upon, and appropriating to their own use, parts of the line belonging to the Rail Road." It would not be until the fall of 1855 that a "Referee Award" would decide the outcome of the Plank Road location in question.

Again, we see the area continuing to grow. In the June 24th edition of the Lake Superior Journal we find that "business of every description is active." Note is made that a lot of iron ore "lies at the Bay awaiting shipment" and then notes that much of it may have to wait until the summer of 1855 when the "Soo" Locks finally open. Note is made in the same article that, for the first time, "considerable ground has been cleared and put into crops." Up to now much hay and food has been imported by boat. The paper also notes that "the influx of population this spring, has been large, and is steadily increasing."

Mr. Ely is also involved in seeing a wagon road built south from Marquette to Bay de Noquet on Lake Michigan. Note is made that it is going through dense forest, but that 40 of the 57 miles will be completed by fall of the 1854 year. The article ends: "the road will be the great thoroughfare of travel, from the mining region north and south in the winter, and will furnish facilities for travel which the Upper Peninsula has never had before." (21, 6.24.1854)

We go now to the next map. On August 26, 1854, the Register of Deeds Office showed that permission was made for the Plank Road to lay its road or tracks, through the Front Street area of Marquette. Going down towards the Lake below Washington Street, "on the grade of said Main street for the passage of cars, engines, or other vehicles anywhere between the sidewalk on the south side of the street and 40 feet north of the north line of Block 18, 19, 20, and across Third Street."

Since the Cleveland Co. was also building a dock, they are given the "right to lay such a track on the north side of Main Street..." and "the same shall be laid within 30 feet of the southline of blocks 15, 16, 17." Permission was also given to cross, on, or over Water Street. (The original grade seems to have gone through the lower Front Street cut to its dock as in the photo on page 50. It must have later built a dock further south by Ripley's Rock, and crossed Spring Street.)

Permission was also given to the Plank Road and the Rail Road Co. the right to cross Third, Front, Washington, and Water

Street with their excavation to the docks being constructed. There were two excavations through Front Street, and wood bridges were built to pass over the Railway cuts.

The map is a later one and shows the Cleveland Mine Dock serviced by the Rail Road and crossing Spring Street. The Lake Superior Mine will build a dock between them with another cut through Front Street. The cut, but not the dock is shown here.

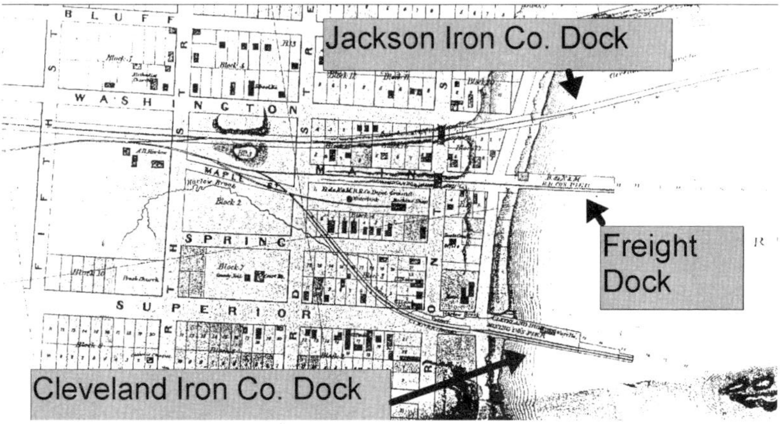

On August 26, 1854, the newpaper noted that a request from the Carp River and Iron Bay Plank Road Co had asked for a dissolving of the injunction against them by the Rail Road, but it was denied. The injunction, of course, had stopped construction of the Plank Road at several places. The Rail Road would see the Plank Road moved in several places in 1855.

In October of 1854, a large storm occurred on Lake Superior and the Sharon Iron Co. lost their $15,000 dock as mentioned previously. Note was also made in the newpaper on October 7, 1854 that the "Iron Mountains" (eventually Ishpeming and Negaunee) "are completely blockaded by fallen trees."

Whether it was the loss of the dock, or not, the Sharon Company left the Upper Peninsula. The Jackson Mine and Forge returned to their original names, and the new buyer, the Clinton Iron Co., only lasted a year. They gave up in 1855 and Peter

White and J. P. Pendill (of Negaunee) took over for a few months and they, also, gave up any more attempts to run the Forge. Like the Marquette Forge, the Jackson Forge would not ever get to use the new Iron Mountain Railway or the Rail Road.

In the winter of 1854, the final act needed for the opening of the Plank Road was completed. It was the digging out of the part of Marquette's Front Street to make the Jackson Cut. With the opening of the Sault Canal in early 1855, the Jackson and Cleveland Mountains would be ready to ship lots of ore on their Plank Road.

Pictured is the Mineral Rock propeller ship of about 1855. From the Ralph Williams' book, p. 143, through the J. M. Longyear Research Library in Marquette, Michigan.

This is a photo of the old Plank Road which became the M-28 highway between the Negaunee city limits and going east, between the original city cemetery and the present city cemetery. It was abandoned about 1970. Note the downhill grade for mules pulling loaded ore wagons.

This is a photo of the Plank Road just east of the Deadman's Curve area going towards Marquette.

Chapter 6
The Year 1855

The first main event to be in the newspapers in 1855 was neither the Plank Road or the Rail Road. It was the opening of the Sault Ste. Marie Canal. The first shipment of Iron Ore through the canal was made by the Cleveland Iron Company in Ishpeming. The bill of lading reads as follows:

"Marquette, Mich., August 14, 1855. Shipped in good order on board the brig Columbia, Wells master, by the Cleveland Iron Mining Co. the following articles consigned as per margin." "For Cleveland Iron Mining Co., Care of Crawford and Price, Cleveland, Ohio. Unless otherwise ordered by W. J. Gordon of Cleveland, Ohio. One hundred tons iron ore $2.75 = $275.00." and signed "J. Wells." (31, p. 136)

Note is made that in 1855 the Cleveland shipped 1.447 total tons of iron ore and that the Jackson and Lake Superior did not ship any ore that year. (32, p. 137) The ore from the Cleveland Mine had to be sent on the original dirt Wagon Road, the Plank Road still being incomplete in August of 1855. The Jackson Mine may have been waiting for the Railway (Plank Road) to be completed in 1856, in spite of the Lock being open at the Sault. The Lake Superior was waiting for the Rail Road to be completed to its mine which would be 1858.

In May of 1855, an official survey was made of the Plank Road, probably for use in the legal battle by the Rail Road. This written survey of exact footages, and exact angles has been printed in the back of this book as Appendix "B." This written survey seems to be the final route of the Rail Road after the Plank Road did its moving.

Another large problem that existed before the Plank Road, and will continue to exist after it was completed, was the slow loading process at the docks. Ralph William's book quotes Peter

White and notes that the pig iron from the furnaces in Marquette County were all loaded by manual labor and "at a cost per ton that would today be absolutely fabulous." It is noted that it often took four days to load a boat with 300 tons of blooms or pig iron using block and tackles (with pulleys), and a horse. He gives us the details as follows: "A block carrying manila rope was fastened to the ship's rigging and a tub attached to the rope...and lowered into the hold. Another block through wiich the rope ran was fastened to the dock and the horse hitched to the end of this rope. The tub was filled by hand shoveling and then hoisted by the horse walking forward. To get the tub back into the hold it was necessary for the horse to back up." (32, p. 142) And it was noted also that it was even more difficult to get the iron out of the ships hold upon reaching down-lake ports, and that often the schooners preferred to carry the entire load on the top deck rather than down in the hold where it had to be lifted twice to reach the dock.

A very large event took place in Marquette in the summer of 1855. The "Brig" Columbia made its appearance in the Marquette Harbor at two in the afternoon on August 7. It carried the first locomotive, called the Sevastopol, or the Sebastopol as it became both in spelling and pronounciation. (In 1921, F. H. Begole presented the next photo, claimed to be the Sebastapol. He noted that the E. C. Robert's brought it up the Lake. I believe he means that it is the Heman Ely, the second locomotive, which was brought by the E. C. Roberts vessel. The photo is from 1901 when Lake Shore Engine Works repaired it. The photo really does not look like either of the first two engines, but might be one of the first four used on the I.M. RR. Or it might be that the smokestack could have been lengthened later and other changes made in the 50 years since its arrival in 1855. A sketch of the Sebastopol will follow the photo.

The Lake Superior Journal recorded the locomotive's arrival in their August 18th issue, as follows: "As soon as the brig came inside the pier, a gun (made of Lake Superior iron) was fired as a signal of welcome to the Iron Horse and tender, to which the bell of the locomotive responded, "I am coming, etc." Salute after salute was fired and responded to, until the brig Columbia let go her anchor in front of the dock and town. At this moment a gun was fired that made the hills and rock and iron fairly tremble, the bell of the locomotive responding with louder peals than ever, "I have come, etc., etc.."

The newspaper on August 11th also noted that a crowd of "Natives" were among those gathered to see the "iron horse" as it arrived. All present were evidently sharing the stories of the "monster's feats in traveling and its capacity for carriage." The paper also noted that "The Iron Horse has now become one of the fixtures of the Lake Superior region, and the time is not far behind when the shrill neigh of the tireless steed will be heard echoing through the whole of the Iron and Copper coast."

The Sebastopol was put right to work in the building of the Railroad. It was noted in the September 22 paper that it was making "several trips per day with ties, rail, etc. It was expected to reach the Eureka Mining Co. location within a few days, and by means of connecting it with a branch road that company hopes to ship some 1,500 tons of ore to its rolling mill at Wyandott on the Detroit River this fall."

As you can see, this drawing of the Sebastopol is somewhat different than the photo. It was fairly good size. The total length as shown is just under 52 feet, and the total weight was 94,700 pounds, or about 47 tons with the tender loaded. The steam cylinders were 16 inches, and the stroke was 22 inches. Weight of the engine alone was 31 tons. It was noted in the September 13, 1856 newspaper that it was "not gaudily ornamented."

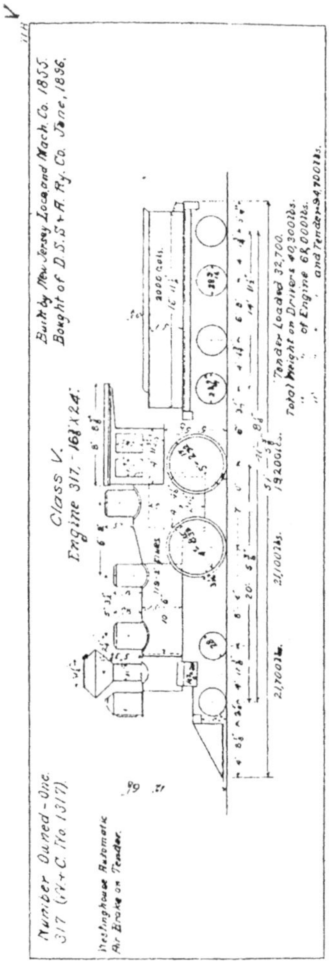

The Sebastopol was built by the New Jersey Locomotive and Machine Co. in Patterson, N. J. in 1855, and surprisingly was still running when the railroad was purchased by the Duluth, South Shore, and Atlantic Ry. Co. in 1896. It is also noted that the Tender was equipped with Westinghouse Automatic Air Brakes. Later in its life it was named the Pioneer, No. 20. (Sketch from Midwest Railroader, No. 40)

The Eureka Iron Mine intrigued me enough to spend a separate day researching it further. The Lake Superior Journal by 1871 had moved to Marquette, and become the Mining Journal. In the June 17 issue, it was reported that a Dr. Hewitt had a force of men at work at the old Eureka Mine, 1.5 miles from Marquette, working a narrow vein of ore and some of it was hauled to the Rail Road track. In July of 1872 it was noted that the Eureka vein had grown somewhat narrower as the ore was removed. Three years later in 1875 it was noted that "The old Eureka Mine, so long abandoned," has twenty men employed and is taking out an excellent quality of Hematite which was being taken to the Bancroft Furnace for purifying into pig iron. The Bancroft was built on the site of the old Forest Forge.

On May 22, of 1875, the mine was renamed the Iron Bay Mine. It was noted that 24,000 tons had been mined and it was making first class iron at the Bancroft Furnace in Forestville. Note is also made that the iron is found in narrow veins with soapstone.

In July of the same year a 48 foot deep shaft was dug., At the bottom was discovered a six-foot vein of ore. Yet this note was also given in the same article: "Today the mine could scarcely be given away."

A last article, in October of 1875, noted that the Eureka Mine, two miles from the City is to start operations in one shaft with a contract with Morgan Furnace to buy the ore. Note is made that shaft must have been full of water at this time because a "horse-power whim" has been rigged up to empty the shaft in twenty hours. A Mr. Ward is looking for miners to mine the vein at the shaft bottom.

In Mr. Wesley Perron's notes I find this very good information from 1906: "Section 21, T48N R25W, Harlow's Saw Mill built by Amos R. Harlow one of Marquette's earliest pioneers. (This was also the location of the Eureka Mining Co.'s mine, NE quarter of the NW quarter, same section, town and range....) My calculations put it on the east end of the Cox Property Map in Section II of this book..

The final word is in an August 10, Daily Mining Journal of 1929 (or 54 years later). Believe it or not, the mineral vein under the bottom of the shaft in the Old Eureka Mine is still being investigated in 1929 by the Cleveland-Cliffs Iron Company. The location of the mine is now given as a short distance from the end of West Washington Street. A diamond drill is being used for two weeks and may take a month more to find the results.

The drill will go down a distance of about 500 feet at the most. The CCI felt that the decomposed iron hematite might contain pyrites and other metals. It was noted that ores with pyrites are valuable in furnace operations. The property description states that it as a quarter section of land originally purchased in 1853 from A. R. Harlow. The final written location of the mine in the 1929 paper was given as: "just outside the city limits, east of Vandenboom's farm and south of the road which runs to that farm from the end of West Washington Street." I have not located the mine. I leave the mystery to the reader.

Let us return to the year, 1855, and let us see what the Iron Mountain Railway (Plank Road) is doing by September 22. The newspaper notes that the Plank Road is about finished except for the laying of the rails, and that is being done at about the rate of a half mile per day and is finished up to "Riplins" (unknown). It is noted that both the cars for carrying ore and the mining gear for the Plank Road have arrived, and beginning operations are thought to be starting "the first of October." Of course, the article then goes on to mention about the problem of the Railway and the Rail Road both trying to get through the narrow Gorge, not too far from present U.S.-41 and the LS&I RR overpass. That Award, made by C. T. Harvey, came out in October, requiring changes to be made by the Plank Road. Some of these details were listed in Chapter Three.

We learn, regarding the suits, that both parties had agreed (dated September 13, 1855) in a friendly manner to allow Charles T. Harvey, Esq. to decide if both roads can use the desired route, and have authorized Mr. Harvey to relocate the roads at "his discretion" and without "material detriment to either." Mr. Harvey accepted the task.

In October, 1855, Mr. Harvey gave his decisions to both the Railway and the Rail Road. Although the survey map covered "Stations 14.04 to 743.0, or the entire fourteen mile route, the Referee Award covers only the problems between Station 191 to 550.44. Station 191 starts in the area between Wright Street and Grove Street in Marquette, and ends at about the present Negaunee Cemetery. The Plank Road before that area and after that area must have continued in construction during 1855. We do know that after the Referee's Award on October 2nd of 1855 that the Iron Mountain Railway (Plank Road) had to spend some time blasting rock and moving its planks as directed to do by the "Award." According to one writer in 1909, the Plank Road went into operation on November 1, 1855. (31) Others say the date was 1856. Perhaps, although November, 1855, may be a completion date, it may have been too late for the Plank Road to do any shipping.

The Rail Road is also moving along, and is completely completed past the Eureka Mine and is building a switch from the mine to connect with the Rail Road. The paper notes that the mine at this date is blasting out ore in considerable quantities and that "as they go down the ore appears to improve in quality and quantity."

The final article for the year 1855 is in the December 1 issue of the Lake Superior Journal. There is no note that the Plank Road has been completed, but it is noted that the Collins, Forest, and Jackson Forges, although shipping 2,000 tons, might have shipped much more if the new facilities had been in existence. Facts and figures are estimated for all three Iron Mountain mines for the coming year, with a total of expected shipping to increase a hundred-fold, to 200,000 tons by the Plank Road Railway. The article notes that the Rail Road is also expected to be running by 1856 (It will be partially running in 1857). A blast furnace is expected to be built in 1856 as well, near the mouth of the Carp River, built by the Peninsular Iron Co. The year 1856 looks like a very exciting one, remembering that the Sault Ste. Marie Canal is now operating and waiting for great ore tonnage.

We should also add another very important improvement for our area. In 1855 Mr. Everett wrote to a friend, "The long talked-of road from here to Bay de Noquette is now opened, and the first team arrived here yesterday with sixty bushels of grain, the first load of anything ever brought by the overland route to Lake Superior." Note was also made that up to now, only dog teams arrived in the winter from Green Bay with mail. Now mail will be expected once in ten days. (27, p. 105)

Here is the Iron Mountain Railroad heading down into the Caved-in area of Negaunee to the left. Note the fence in the foreground.

This is another view of the Plank Road and the Iron Mountain Rail Road as they come south to present County Rd. 492 from the area of the later Morgan Furnace to the present Eagle Mills area.

Chapter 7

1856: The Plank Road is in Use, but the Rail Road has Problems

Frank Stone, in his book, notes that with the Sault Canal opened, a load of up to 50 mule teams were brought to Marquette for use on the Plank Road. He tells us that half of the mules were at the Marquette end, and half at the Cleveland and Jackson Mines. (28, p.103) At this time, six-mule teams had to haul raw ore down to one of the Forges or Furnaces at or near Marquette, and then return. Eagle Mills was the meeting place for passing.

Harlan Hatcher's book tells us that the Plank Road had steep grades and it was hard to brake and hold back the heavy cars. "Too often carts got out of control and rolled down on the mule teams, crushing or killing the expensive animals." In another article we also learn that the four-wheel wooden carts drawn by mules made one trip per day with a single car of one and a quarter ton of iron ore in each. (4)

A peculiar twist in facts, which seem to be totally unbelievable, came from a Michigan Historical Magazine, volume 28, of April-June, 1944: According to the article, the Plank Road ran without rails in 1856, but in 1857 it was changed to a primitive railroad with primitive wood-burning steam locomotives with little wooden "jimmy" railway ore-cars carrying four to six tons of ore. All of the plank road ties lying next to each other as a pavement were removed and regular spaced ties were laid down. It notes the track was made of hardwood lumber 4 inches by 6 inches by 16 feet long. The strap irons were 3/8 of an inch thick and 2 inches wide, and bolted on with spikes. It was noted that the spiked ends would often come off and spring up through the floor of passenger cars, with damage and danger. The article has this railroad running yet in 1861 but on then light iron rails.

I read nothing of any of this in the U.P newspapers, and it almost seems to be the story of a different Plank Road. We know already that the large engine Sebastopol will be operating in late 1857 from Marquette to Ishpeming on Iron T rails, and then continue on to National Mine and points west.

One early book of 1907 does seem to have a real insight into the problems with the Plank Road. It was the cars that hauled the ore. A Dr. Hewitt notes that for two years they tried to haul 4 tons each. He noted that "the cars were flat bottom affairs and when they were first loaded it was found that the load bore so heavily upon the tracks that the wheels scraped against the bottom of the car, which was remedied by cutting away a portion of the platform just above the wheels." The 1907 book also notes that a team could make only one trip a day, and it took all teams to get 35 tons from the mines to the Lake in one day. "The grades were frightful and the cars frequently ran away, mangling the mules and jumping the track at the first curve." Note is made that the mules had cost $1,400. per pair. It was also noted, and the following sounds like a very serious problem, that the tracks frequently got out of shape. (32, p. 144)

Thus, already in 1856, the Cleveland and Jackson Companies soon saw the need for the regular railroad and the use of the Sebastopol for carrying their ore. They sought consolidation of their Iron Mountain Railway with the Iron Mountain Rail Road.

And then Mr. Heman Ely died, suddenly died, and was buried in Park Cemetery in Marquette. His brother came from New York and carried on the work of the Rail Road. His name was Samuel Ely, and although not mentioned much in this book, did a great many things for Marquette County. The other brother, George, who helped Heman, then aided Samuel.

By May 17,1856, the population of Marquette and surrounding area grew to a thousand. Two churches were being erected, the Episcopal and the Methodist. Fences were going up in front of homes, and plank sidewalks being built to avoid walking on muddy and dusty roads. All of the lumber mills were unable to produce enough wood. People are also purchasing cows.

The Plank Road, according to the May 17, 1856, paper, is running smoothly, and a second article in the same paper says the Cleveland Iron Mining Co. is using 26 mules, and four horses as well, on the Plank Railway.

The Rail Road was completed for 3.5 miles in May for locomotive use. The Eureka Mine is anxious to use it. Although they are having a difficult time getting the ore out, they feel the short distance to the dock will save them money over the Iron Mountain Mines that are 17 miles up the line. There is still skepticism about what ore is there, but some believe it may be "an Eldorado" (A legendary place of untold riches).

The Rail Road problems mainly were some soft spots in the grading. The grading was going well on both ends, but not in the middle. Three miles not graded were "about half way between the Lake Superior and the Cleveland Mountain." (July 26, 1856). This was the swamp, no doubt, just south-west of the gorge.

The Rail Road is also building several new buildings. The details of these buildings near the Lake are to be found in Appendix "C." We will comment here on the Sebastopol as we learn in the paper on September 13, 1856, that the engineer is Mr. Donkersley. It is noted that he is an excellent workman on the engine as he was formerly in the "employ of the New Jersey Locomotive and Machine Co, and helped to build the Sebastopol. He also returned to the factory to oversee the building of the second engine. The New Jersey Co. was the builder of the eventual four engines the Rail Road purchased.

As the winter of 1856 arrived, in spite of a road now coming from Lake Michigan's Bay de Noquet to the Marquette area, large tonnage of goods and large items still had to come by boat before the harbor froze over. There were still fears causing worry about winter and facing the cold if goods did not arrive by boat. The following information comes to us from the book, The

Honorable Peter White: A newly built, large boat was to arrive in Marquette in November of 1856, and was to be carrying a cargo of "feed for the horses, and provisions for the men. The feed for the horses is mentioned first because it was the most important item, as there was scarcely a ton of hay left in the village, nor was there any means of obtaining any until spring." Note is also made that it carried needed mining supplies. "Unfortunately, the Webb was burned on her first trip in the shelter of Whiskey Bay, down to the water's edge, and sank, causing the destruction of her entire cargo."

A disaster was only prevented because a Mr. J. Tallman Whiting decided to duplicate the cargo on a propeller boat and he left the lower lake late in the year. The ship luckily not only made it to Marquette with the needed supplies, but went on to Ontonagon where it spent the winter. (32, p. 144)

This is a photo of the docks in Marquette in 1857, courtesy of the Marquette County Historical Society and the Ruth S. Main estate. Docks appear to be higher than four feet above the water's edge. The top-most dock must be the Cleveland Mine's at this time, along with their warehouse. The Jackson Mine Dock is the lower dock in photo. Both docks were being served by the Plank Road (The Iron Mountain Railway) at this time.

Chapter 8

The Completion of the Iron Mountain Rail Road

1857. On April 25, the newspaper speaks of the completion of Mr. Ely's Rail Road Dock, and notes that the trestle to it would soon be completed. The Cleveland Mine is making an extension to its dock and will build the trestle to it when ice leaves the harbor. The ditch across Front Street had been completed, and the dirt used for the dock approach. The Rail Road would also build the Front Street wood Bridges.

A second locomotive also arrived. The Ely's had contacted Philo Marshall Everett, who was wintering in Cleveland, to find a vessel to bring up the Heman Ely. Everett found the E. C. Roberts, which reached Marquette in June. The vessel brought the locomotive, and iron T rail, and railroad spikes. A total of 390 tons was delivered. Note is made that the vessel still had to go through quite a bit of ice at the Marquette Harbor on June 20. Mr. Everett notes that year in his diary: "The last piece of ice was seen floating in the bay on the morning of the third day of July, 1857." (28, p. 109)

It only took a few days to havethe "Heman Ely" running. It was used on the part of the Rail Road already built to add ballast to the ties and smooth it up. Note was made in the July 25, 1857, Lake Superior Journal that "All Marquette, and his wife, ought to make a point of visiting this new machine for it is one of the finest in the country." It is noted that Mr. Donkersley was at the New Jersey works where it was built and "spent the winter in overlooking its building."

We give you here the sketch of Locomotive No. 2. One of the saddest things that took place at this time, is that Mr. Heman Ely, who had done all the work in heading up this Rail Road and started several other projects, but who had died the year before, had his name removed from his own engine, and it was renamed after Mr. Donkersley from then on.

Details for this new locomotive are: Weight of the engine only: 26 tons. Cylinders are 16 inches with a 24-inch Stroke. Flues for steam are 137 inches of two inch diameter. Driving wheels are 4.5 feet and weight on the drivers is 16 tons. Total running weight of engine and tender is 51 tons. Running boiler pressure is 130 lbs, and the length is 53.5 feet. It was very ornamental, constructed of brass "which shines like gold in the sun." It was also noted that "the painting and gilding are of the highest order of this class of work" Also noted are a magnificent head light, and patent brakes. It was built as a freight engine and had a longer wheel-base than the Sebastopol which was made as a passenger engine. (August15,.1857) (Note that some figures vary slightly from those on the sketch.)

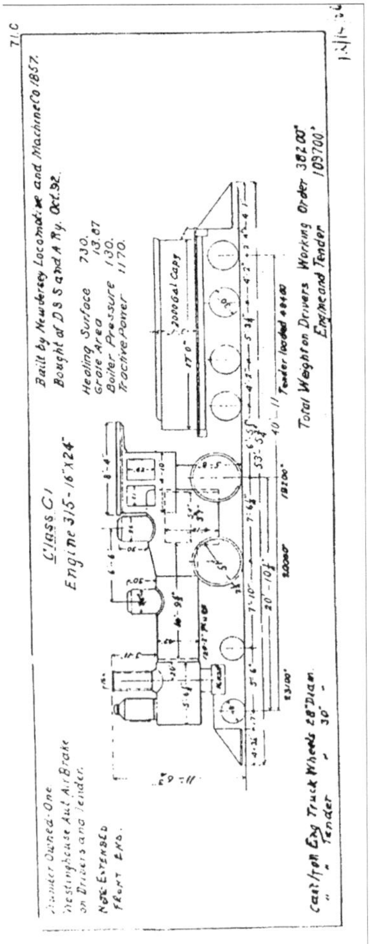

In July, the paper notes that the Sebastopol locomotive went on rails laid in a wet place and that "the track settled to one side and threw the locomotive off. No injury sustained except the delay of a day or two in getting things to right and repairing the breech, and everything is in good running order again."

Another construction problem for the Rail Road was the swamp it had to pass through after the narrow Gorge with the Railway on its north side. The July 11th paper notes that just west of the gorge "lies a deep swamp hold which has heretofore defied all attempts to lay a track across it. Three times was the grade brought up to level and each time it settled down below the surface. But the fourth attempt has been more successful, and yesterday the locomotive with a train of cars passed over it without disturbing it in the least. The last filling is entirely of stone." Here is that area as shown between the later DSS&A (on left)and LS&I (on right) grades, with the Plank ane Rail Roads in the swamp between:

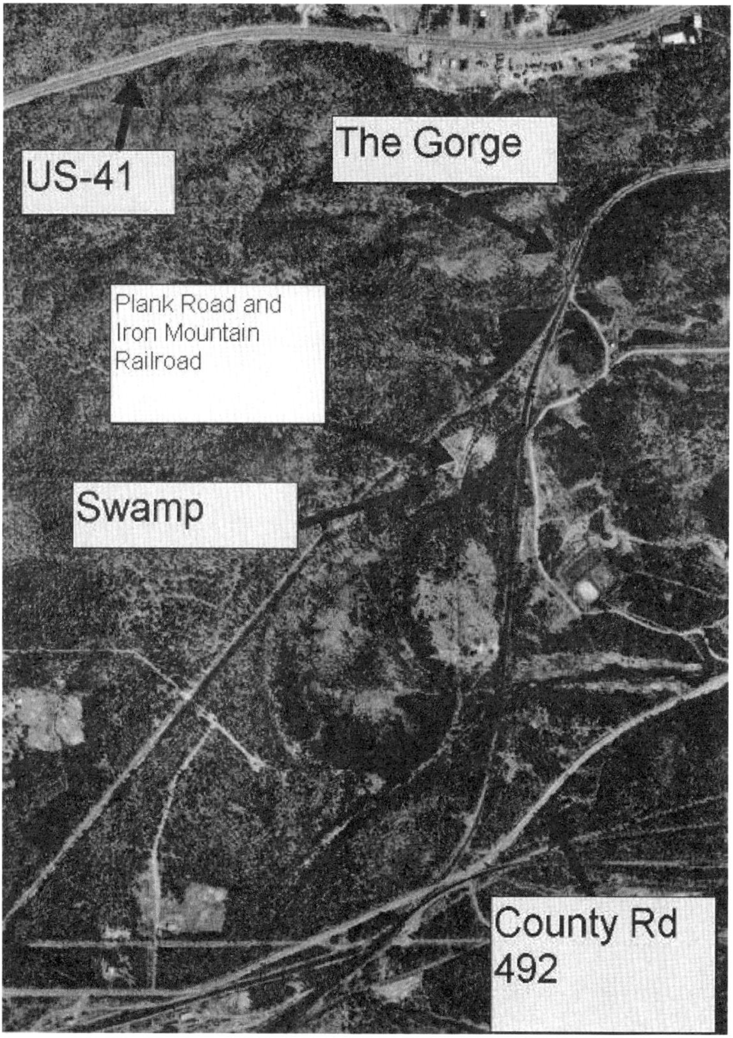

(Photo from Adventure Station.com Reno, NV. And USGS, National Geographic, and offroute.com, and Terraserver.com)

According to an earlier newspaper of April 2, 1857, the Plank Road was already going through this area just fine and had already conveyed 18,000 tons of 75 per cent iron ore.

Then a bad set-back occurred that affected the Rail Road and the entire iron mining area. The book on Peter White notes that 1857 was a bad year for selling iron ore to the lower lakes' furnaces. "When things are flat, iron is the flattest of them all." Money, actual currency, disappeared from wallets, from stores, and from banks. Locally, a Mr. W. J. Gordon came up with an idea for "Iron Money." It was not coins, but paper notes of small denominations issued by the iron companies. Local banks readily accepted them. Banks saved them up, sent them to the Mine offices, and they were paid with interest. The "Iron Money" was still in use locally up to 1872. (32, p. 151)

In 1857, The Pioneer Co., of New York, started to build the Famous Pioneer Furnace in Negaunee, with the Iron Mountain Rail Road building their grade to the property. The Cleveland Mine's Marquette Furnace, purchased from Mr. Graveraet, had burned down and the newspaper announced that the Cleveland Co. entered into an agreement in 1866 with the Pioneer Co. to run the Pioneer Furnace, giving the New York owners 30% of the profits. Eventually the Cleveland Mine purchased the furnace, which had two stacks. It was noted that a village was growing up around the Pioneer Furnace. (23, 9.29.1855)

In 1857, area men felt that the community growing around the Jackson Mine and Pioneer Furnace needed a name, and felt that Indian names were the most suitable. A meeting with area Ojibwa Indians chose "Negaunee" meaning "the first, or pioneer." In 1858, Samuel Ely, Mr. White and a committee chose "Ishpeming" meaning "remarkable elevation" because water there goes both to Lake Michigan and to Lake Superior. (31, p.158)

The July 18th newspaper informed the readers that the Iron Mountain Rail Road is now up to the Carp River (near the present Negaunee Cemetery and the Plank Road 492, but south of it), and nearly completed across the "great Carp Marsh." The engine cannot run on it yet, and even teams of animals were impossible to use on the "soft, peaty surface." That meant that men had to carry each tie in and lay it down. Still, the newspaper notes that the Sebastopol should be at Jackson Mountain by August 15th. It was another of those useless predictions.

In August, the Rail Road was running up to Duncan's Steam Mill, nine miles from the harbor. This is where Eagle Mills is today. The Cleveland Mine in Ishpeming decided to use the Rail Road from there, down to the Lake. Using horses on the Plank Road (Iron Mountain Railway—also now called the Strap Rail), they pulled cars of ore to Duncan's. The ore was then put on the iron T's of the Rail Road to be taken by locomotive to the docks. The Rail Road was worried that the Plank Road cars would not hold together, and as soon as the new regular Rail Road Cars arrived, they were assembled, and used to haul the Cleveland Ore. (July 22 newspaper)

This leaves us with several questions. Were the rails of both Roads the same width? We know that the Iron Mountain Rail Road space between T rails was 4 foot 10 inches.(2) (How it got to be 4 foot 8.5 inches on today's railroads is another good question.) If the Cleveland cars were narrower, and ran between the Rail Road rails, it must have been quite a jolt between each tie, unlike the solid row of the Planks. And if the Rail Road was to use its own cars from Duncan's downhill, did each Cleveland car have to be unloaded and reloaded? We do know that the Cleveland Mine was able to send twice as much ore each day by using the Rail Road from Duncan's, rather than using the Plank Road all the way. (August 22, 1857)

The newspaper on August 29, 1857, notes that the Iron Mountain Rail Road track is now built almost to the Pioneer Works. However, track is also being laid from the Jackson Mine to the Cleveland Mine. This part was easy in one way, and more difficult in another. The grade was all ready there, because the Rail Road grade would use the Plank Road grade. But it was more difficult in that all the wooden rails of the Plank had to be removed, and the iron T rails installed. No mention is made as to whether the Planks themselves stayed as the Rail Road ties.

The Jackson Company also began using the new Rail Road, but was transferring their ore about three quarters of a mile more towards Negaunee from Duncan's Mill. We find it is the Donkersley Locomotive now hauling the ore to the dock, and the Sebastopol is doing the ballasting of the ties, and that the Carp River area still needs ballasting..

The same newspaper notes that there is a fleet of boats waiting to be loaded, and that 200 tons of ore a day are being brought to the docks. And it is correctly noted that "it will be next season before we can see the full and legitimate results of the road." (Rail Road)

This most interesting note is found in the August 19, 1857, Lake Superior Journal: "When all the companies have put up trestle work on the piers, and fitted up hoppers and shoots (chutes), the amount of handling of ore in loading vessels will be quite trifling..." Note that the modern form of the way we still load ore boats today was already in the minds of men before the first Rail Road was fully completed. It is noted in the book by Ralph Williams that it was the Cleveland Company that was the first to adopt the pocket system of loading ore. In its construction of the dock in Marquette in 1858, they started with about ten pockets with chute attachments. "The idea came from the coal fields of Pennsylvania where coal was handled by means of a chute." (32, p.148)

The October 5 paper notes that a storm brought down several trees on the Rail Road line and that one was hit by a train in the "gorge" with one car being pitched off of the track. In doing so, it lost its ore, but was able to be put back on the track and did not suffer any damage. I have wondered as I walked the gorge area (now filled to a quite high grade) if any of that ore, especially if it is in pig-iron ingots, was still laying there to be found.

The October paper article also noted that the Rail Road is now going to the Cleveland Mountain, but has to be completed 3,000 more feet to the Peninsula Iron Mountain, which became the great Lake Superior Mine.

The area is growing so fast, that in another seven years, 1864, the Chicago and Northwestern Railroad will be in Negaunee from Chicago. It will gain a lot of the business of the mines as it will ship from Escanaba, Michigan and have a longer shipping season than will the Marquette port. (7)

APPENDIX 309

Date	*From*	*To*	*Miles*	*Constructed by*
1864	So. Branch Jct., Chicago	Ogden Ave., Chicago	1.75	Chicago and North Western Railway Company
1864	Escanaba, Mich.	Negaunee, Mich.	62.00	The Peninsula Rail Road Company of Michigan
1864	Magnolia, Wis.	Madison, Wis.	31.80	Beloit and Madison Rail Road Company
1864	Marshall, Iowa	Nevada, Iowa	29.00	Cedar Rapids and Missouri River Railroad
1864	Winona, Minn.	Rochester, Minn.	49.50	Winona and Saint Peter Railroad Company
1865	Little Rock Island	Clinton, Iowa	.22	Chicago and North Western Railway Company
1865	Nevada, Iowa	Boone, Iowa	23.00	Cedar Rapids and Missouri River Railroad
1867	Boone, Iowa	Missouri River, Iowa	149.60	C R & M R R.R.
1867	Rochester, Minn.	Waseca, Minn.	55.50	Winona and Saint Peter Railroad Company
1867	Missouri Valley, Iowa	California Jct., Iowa	5.84	Cedar Rapids and Missouri River Railroad
1867	California Jct., Iowa	Sloan, Iowa	49.81	Sioux City and Pacific Rail Road Company
1868	Sloan, Iowa	Sioux City, Iowa	20.00	S C & P R.R. Co.
1868	Glenbeulah, Wis.	Fond du Lac, Wis.	23.40	Sheboygan and Fond du Lac Rail Road Company
1869	California Jct., Iowa	Fremont, Neb.	31.77	Sioux City and Pacific Rail Road Company
1869	Fremont, Neb.	Maple Creek, Neb.	10.00	Fremont, Elkhorn and Missouri Valley Rail Road Company
1870	Negaunee, Mich.	Lake Angeline, Mich.	4.58	Chicago and North Western Railway Company
1870	Various branches to Michigan mines		42.27	C & N W Ry. Co.
1870	Winona Jct., Wis.	Winona, Minn.	29.00	La Crosse, Trempealeau and Prescott Railroad Company

From the book, Pioneer Railroad,
The history of the Chicago and Northwestern Railroad, listed as #7 in the bibliography.

The Jackson Cut

I am standing not too far from where the Jackson Mine Dock was located (behind me) and looking west. Note the triangular building on the right allowing for the angled cut for the Plank Road and the Rail Road under Front Street. The bank in the background has since expanded into the Plank Road space.

Chapter 9

The Rail Road Begins; The Plank Road Becomes the Highway

A new era began with the completion of the Iron Mountain Rail Road. By 1858, it was totally successful. One can hardly believe the large increase in the amount of iron ore going to the ships, let alone passenger traffic, and general freight. Large posters were printed, the top of one which is printed below:

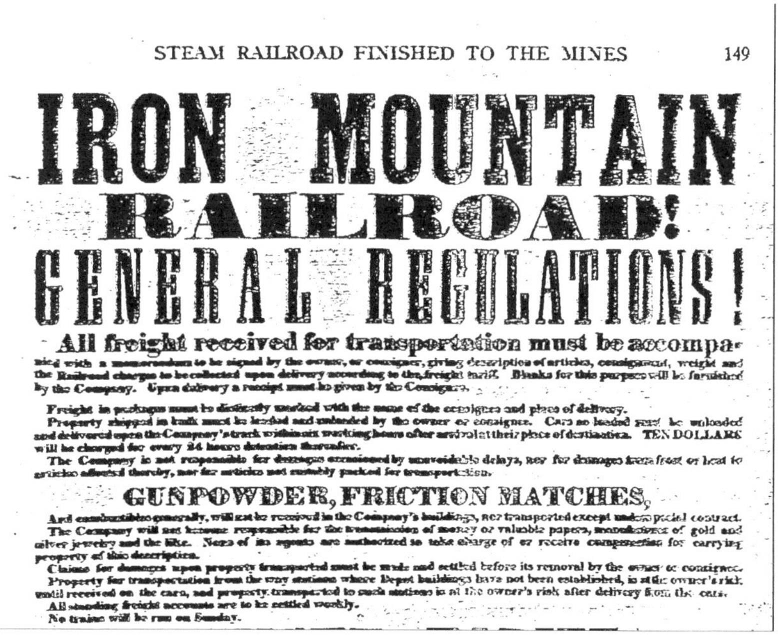

STEAM RAILROAD FINISHED TO THE MINES 149

IRON MOUNTAIN

RAILROAD!

GENERAL REGULATIONS!

All freight received for transportation must be accompanied with a memorandum to be signed by the owner, or consigner, giving description of articles, consignment, weight and the Railroad charges to be collected upon delivery according to the freight tariff. Blanks for this purpose will be furnished by the Company. Upon delivery a receipt must be given by the Consignee.

Freight in packages must be distinctly marked with the name of the consignee and place of delivery.

Property shipped in bulk must be loaded and unloaded by the owner or consignee. Cars so loaded must be unloaded and delivered upon the Company's track within six working hours after arrival at their place of destination. TEN DOLLARS will be charged for every 24 hours detention thereafter.

The Company is not responsible for damages occasioned by unavoidable delays, nor for damages from frost or heat to articles affected thereby, nor for articles not suitably packed for transportation.

GUNPOWDER, FRICTION MATCHES,

And combustibles generally, will not be received in the Company's buildings, nor transported except under special contract.

The Company will not become responsible for the transmission of money or valuable papers, manufactures of gold and silver jewelry and the like. None of its agents are authorized to take charge of or receive compensation for carrying property of this description.

Claims for damages upon property transported must be made and settled before its removal by the owner or consignee.

Property for transportation from the way stations where Depot buildings have not been established, is at the owner's risk until received on the cars, and property transported to such stations is at the owner's risk after delivery from the cars.

All standing freight accounts are to be settled weekly.

No trains will be run on Sunday.

The bottom of the poster listed the Freight Tariffs and passenger charges.

Passenger fares are listed as from Marquette:
To Franklin, .15 to Duncan's Mills, .30

To Jackson Mountain and Pioneer Works, .42
To Cleveland Mountian, .48 To Lake Superior Mtn, .54
The Railroad notes that if people want half-rates, they might put on a "second class" (open?) car.

Lumber will be hauled to the Negaunee area for $1.65 and to the Ishpeming area for $1.85 and $1.95. All are for 1000 board feet (2 x 6). The actual names of Negaunee and Ishpeming are coming into use in 1857 and 1858.

In the last two weeks of August, up to September 4, 1858, the Lake Superior Journal notes that the Rail Road hauled 395 tons of iron ore from the Jackson, 761 tons from the Cleveland, 405 tons from the Lake Superior, and 199 tons from the Pioneer Furnace. The total of the three mines for the shipping season, up to September 4th has been 23,063 tons.

Regarding the above paragraph, if the Rail Road is hauling raw ore, it may be going to a furnace in the Marquette area to be made into pig iron. It is my understanding that no raw ore is shipped by boat at this time.

In 1861, these statistics were printed as being shipped from Ishpeming and Negaunee:

Year	Pig Iron	Iron Ore
1858	1,627 tons	30,556 tons
1859	4,683 tons	83,078 tons
1860	3,560 tons	150,903 tons

Passenger business (note made: No free passage or tickets)

Year	Number	Receipts
1858	4,629	$1,540.62
1859	6,445	$2,007.42
1860	5,487	$1,989.92

Merchandise and Freight

1858	1,806 tons
1859	2,258 tons
1860	2,124 tons

The Iron Mountain Rail Road did not last long with that name. In spite of good business statistics, it was resold briefly to the Chicago, St. Paul & Fond du Lac RR, and was resold again to The Bay de Noquette and Marquette Railroad Company in 1858. That Railroad in turn was consolidated in 1871 with the Marquette and Ontonagon Railway, which became the Marquette, Houghton, and Ontonagon Railroad and then ran as far as L'Anse in 1872. (In L'Anse it also built an ore dock for Baraga County ores.) It was finally purchased by the Duluth, South Shore, and Atlantic Railway Co. in 1887. The Canadians were active in ownership as it provided part of the route across Canada for many years, running from the Sault Ste. Marie to Duluth. Sometime during the 1800's, the original Iron Mountain Rail Road grade that I have walked was moved and rebuilt to the current abandoned Duluth South Shore and Atlantic grade. That grade may become part of the Heritage Trail.

A new railroad from Marquette to the Ishpeming and Negaunee area Iron Mines was the Marquette and Western Railroad Co, which left southwest Marquette and went to the present Marquette Ski Hill Area and then followed the Carp River and Morgan's Creek, crossing the streams over 25 times. The route went in front of the Morgan Heights Tuberculosis Sanitarium, to join the Plank road and Iron Mountain Rail Road grade by the Morgan Furnace. The Furnace, however, was gone by this time already and the original Rail Road grade may have been vacant. The M&W RR was built as far as the Junction Location (Winthrop Junction) south west of Ishpeming in 1884. It later became part of the Marquette, Houghton, & Ontonagon, which used the new route as part of a round trip to Marquette and back, as the M&W track was the least steep of the grades from Marquette. When the DSS&A RR purchased the MH&O, the M&W was thus also purchased. I do not know the years when either the original Iron Mountain Rail Road grade, or the Marquette and Western Railroad grade were abandoned.

The next photo shows part of the M&W trestle where it crosses Morgan Creek just west of the old Morgan Heights Sanitarium. It can be plainly seen from 492 on the north side of the road.

Marquette and Western Railroad Crossing of Morgan Creek, east of Morgan Furnace, as seen from Co. Rd. 492.

Here are some other notes. The passenger trains did not run in those days in the snow. The newspaper noted on December 5, 1862, "The cars having ceased to run a few days since for the winter," that two stages were now operating between the Lake Superior Mine (in Ishpeming) and Marquette. They are run by Mr. Hurley and Mr. S. Kaufman and leave Marquette in the morning and return in the evening.

And....And what has happened to the Plank Road? We know that from the time it was first built that people wanted to use it as a fine smooth highway, albeit up and down every hill. The use of it got more and more popular. In 1858, on February 6, even before the Rail Road was fully running, this notice appeared in the paper:

NOTICE

"All persons wishing to drive teams, or handle freight, on the Plank Railway, before doing so will get a permit from me...."

Also noted was that "none of the Sharon Company's ore cars are allowed to be taken without permission"

It was signed, Samuel Peck.

We don't know what happened to the Plank Railway ore cars of the Cleveland and Sharon Mines, but perhaps they should have been rented for income. We do know that much of the Plank Railway soon became the main road from Negaunee to Marquette as people began to travel it for their personal use. In fact the streets of Negaunee are at the exact angle of the Plank Road leaving town on what became Main Street. The planks must have rotted away, and in about 1912, tarred roads covered the gravel. We are going to spend the second half of this book, actually traveling the Plank Road and looking at the first Rail Road remains. Part of both went through east Negaunee's caving ground (really caved!) area, but much is yet existing.

We do know that the metal straps and the wood rails were removed and salvaged. How do we know? The October 6, 1865, Lake Superior Mining Journal (note the changing name) printed this notice:

STRAP RAILROAD IRON
FOR SALE

About twelve tons, two and a half by five-eighths inches, spike holes punched and countersunk, ends of bars notched and matched.

Price five cents per pound
Delivered at Negaunee or Marquette.
This iron will lay about fifteen hundred feet of track.
Apply to J. Kirkland, Jackson House, Marquette,
Or J. F. Kirkland, Green Bay, Wis.

Note that each bar was notched and holes were countersunk as if two were placed together flatly and one nail used to hold both.

The Iron Mountain Rail Road began a great shipping era. In the first 25 years (to the close of 1884), the Jackson Shipped 2,446,421. tons of ore, the Cleveland Mine shipped 3,024,972.

tons of ore, and the Lake Superior Mine shipped 3,368,560. tons of ore. This must have been raw ore, as a separate figure for the first 25 years for pig iron is listed as 1,100,000. gross tons. The Pig Iron Figure may include mines besides the original three, for surprisingly, we find on the Marquette Range, 72 mines now operating in 1884. (31, p.160)

1868. **1868.**

Bay de N. & M. and M. & O. R.R.

TIME TABLE No. 2

PASSENGER TRAIN TIME

To take effect at 7 o'clock A.M., Friday, May 1st. 1868.

For the Information of Employees only. The Company reserve the right to vary the same as circumstances may require.

	A.M. TRAIN. GOING WEST.	A.M. TRAIN. GOING EAST.	P.M. TRAIN. GOING WEST.	P.M. TRAIN. GOING EAST.
Marquette,	7 00	1 35	5 30	8 25
Bancroft,	7 20	1 20	5 45	8 10
Franklin,	7 30	1 10	5 52	8 02
Gorge,	7 40	1 00	6 00	7 52
Morgan,	7 45	12 52	6 05	7 47
Eagle Mill,	7 55	12 45	6 12	7 40
Negaunee,	8 15	12 25	6 30	7 25
Cleveland,	8 27	12 10	6 40	7 15
Ishpeming,	8 35	12 05	6 45	7 10
Lake Angeline,	8 40	12 00		
Ont. Junction,	8 45	11 55		
Greenwood,	9 10	11 30		
Clarksburg,	9 35	11 10		
Washington,	9 45	11 00		
Champion,	10 15	10 35		

C. DONKERSLEY, Supt.

Courtesy Wesley E. Perron Collection
J. M. Longyear Research Library

PART II
The Walking of the Grade

SECTION I: THE MARQUETTE END

We are going to begin the walking of the grade at the Marquette end. It was here that the Jackson Mine (1848), the Cleveland Mine (1855), and the Lake Superior Mine (1857) built docks to carry their ore. At that time, the docks were built only about four feet above the surface of the water, about the same height as the decks of the sailing and steam vessels. The Jackson was on the north side of the bay, and came from behind Washington Street at an ENE angle to the Bay. It had a deep cut in Front Street, still seen today. The Cleveland built their ore dock almost between the shore and Ripley's Rock. They did not need to cut Front Street because they were at the bottom of the hill at that point. When the Lake Superior Mine built, however, between the other two docks, it used the cut made for the original Cleveland non-ore dock. Front Street thus had two bridges.(9, p. 10) We find from a 1929 article that the F. B. Spear coal dock was built over the Sharon (Jackson) Co. Ore Dock. (5) It is now the Ellwood A. Mattson Park.

The Plank Road, or the Iron Mountain Railway, as it was formally called, only went to The Jackson and Cleveland docks. We will trace the Plank road from there to the City of Negaunee, and point out the deviations of the Rail Road when it was built.

We start with an article from January 1, 1932, entitled: "Old Strap Iron Line is Found." It reads: "While constructing the approaches recently to the new South Shore railway ore dock in

Marquette, men employed by W. G. Miller, Marquette contractor, came upon the old strap iron railway built from Negaunee to Marquette harbor in 1854-1855. The find was made between Third and Fourth streets, near the South Shore main line right of way." The article also notes that it "was located about four feet below the present level of the South Shore railway yards."(2)

The Railway (Plank Road) went "just in the rear of the present buildings fronting on Washington Street," then crossed Washington Street when it reached the bottom end of Cole's Hill (by the present McDonalds). (5) Going north-west it went on the north side of "Lover's Lane" (see arrow) which seems to be about where W. Ridge Street is today. "About one hundred feet to the northward may be seen traces of the old strap iron roadbed" states the 1932 article. (2) It then went westerly and went just to the north of the present County Road (now 492) crossing the present Cox Farm to a point where the road to Forestville leaves the County Road. (5)

The Iron Mountain Rail Road Road "started from Marquette, and the Shops and Terminal buildings were erected on the plot of ground where the Marquette passenger depot now stands. From this point the line was located so as to run through what is now known as the Shop Yards, running westerly and crossing Washington St., near the west end of Cole's Hill near the same place where the Iron Mountain Railway (Plank Road) crossed..." (5)

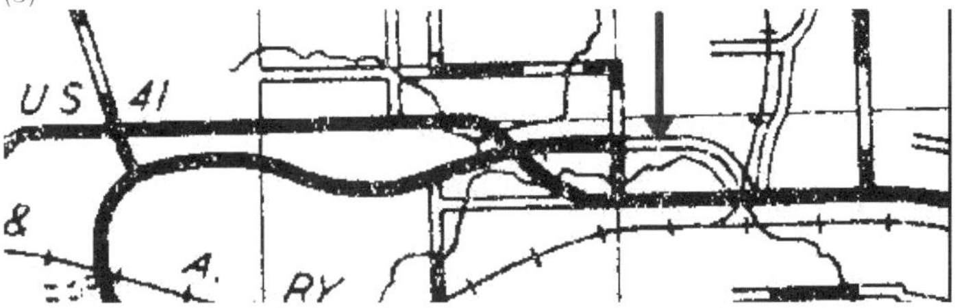

Since one could see the Plank Road to the north from the County Road and since the Rail Road was to the south of the Plank Road the present 492 must lay upon the Rail Road grade. The Railroad grades shown on the above map are of later construction and marked DSS&A RY.

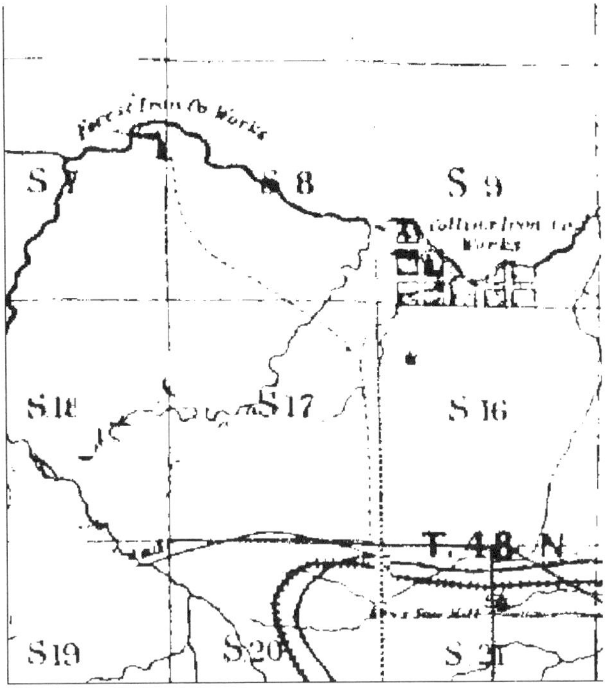

On this map, please note that the Roads to the Forest and Collin's Furnaces were closer to the east Section Line of Section 17 than Wright Street is today. One can see the old road at least in one place. The Forestville Road, where it presently comes south to meet Wright Street, continues south easterly to the top of the hill, and can be visibly seen yet going up in the valley on the east side of Wright Street and north of Commerce Street. See p. 36 for the photo.

After passing the Collins and Forest Road, the Plank Road and the Iron Mountain Rail Road continued on a "long reverse curve." I will call this the "S" curve

I began this book, and walking the Plank Road, using two maps. I divided the maps into portions and walked it in portions. The maps are essentially the same map except one was for the Plank Road and the other reads "The Bay De Noquette (Noc) Rail Road." (The Iron Mountain Rail Road, finished in 1868 became the Bay Do Noc RR the following year.) The big difference between the two maps is the "S" Curve.

The older map which I am calling the Plank Road map has both the Rail Road and the Plank Road, as does the Bay de Noc map, but that S curve is practically absent on the first one.

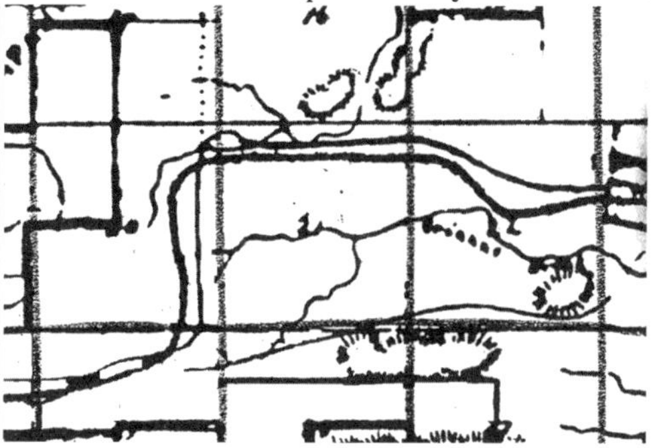

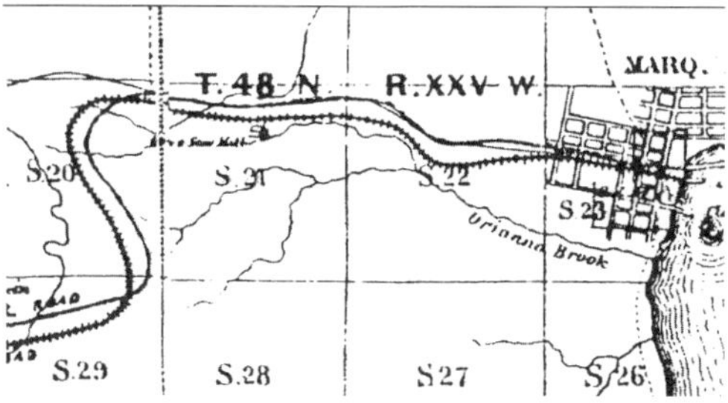

On the above map both the Plank Road (thin line) and the Rail Road (thicker line) are between todays County Highway 492 S curve and closer to the Section Line on the east side of Section 20. I thought I might find them both—or at least the Railroad Grade by walking down the present DSS&A grade from Co. Rd. 492 toward Marquette. After three walks, I found no signs of either the Plank Road or the I. M. Rail Road as found on the top map, and am not sure this first map is accurate. The map below is the accurate one. However, I did find two interesting items.

There is to the right of the entrance into the Lost Creek Housing Development, and before you cross over the stream, an old road down in the woods. It could well be the old Plank Road, being built very flat and about 16 feet across. It goes westerly and stops when you get to 492.

If you continue south on 492, just before crossing the DSS&A grade, you will see on your left, another grade coming out from under the DSS&A. If you look down at the bottom of the valley there, you will see this picture. It turns north by 492 and comes to an end. This may have come from the Furnaces (Collins, Bancroft, and Forest) when this new DSS&A grade was built.

Going over the top of the S curve, County Road 492 switched over from the Rail Road grade to the route of the Plank Road. The Rail Road grade continued westerly, coming from under the present highway on the west. If you walk in on the DSS&A grade to the west but a hundred feet, maybe less, you will see a deep cut on both sides of the track. It is the Iron Mountain Rail Road, and it is at the bottom of the highway grade at this point. As it goes south it goes further from 492 it becomes a high grade and passes a small lake, shown on some maps in this book. The lake cannot be seen from 492, but the grade can be seen some distance away. This must be the part of the Rail Road that was the rough area Peter White spoke about. It has deep cuts, and then a long high grade near the top of this S curve as it goes towards Negaunee and Ishpeming.

On the next page is a photo from 492 of the Rail Road Grade west of the top part of the S Curve. The lake is behind the grade beginning on the right end of the photo.

We see that the Bay de Noquet map is the accurate map and not the original Plank Road and Rail Road map. In fact, the Rail Road grade shown above at the top of the S curve is further west than I thought it would be. It continues to follow the present S curve on the west side of the road going south and staying higher than the highway to the south bend of the curve (bottom end). It is in a very good state of preservation at the lower portion from Bishop Woods Housing Development road, south. Two of the homes of the development on the entrance road are built on the Grade. I park by the Grove Street intersection.

The following aerial map was purchased from Terraserver.com and provided by Adventure Station.com, along with the USGS and National Geographic.

I have added the section line for your assistance in seeing where old Collins and Forest Roads may be, and the placement of the Plank road and I. M. RR, if they did lay closer to the section line. I have also marked where the Rail Road is on the west side of 492. 492 itself is built on the Plank Road, but only to the bottom of the S curve.

The next map below shows an interesting feature of the 1940's, and probably from the 1930's, when the depression travelers needed places to stay in every town: A Marquette State Park. It is on the west center of the S curve
The railroad shown in the bottom left corner is the Marquette and Western (the cross-dashed hump shown).

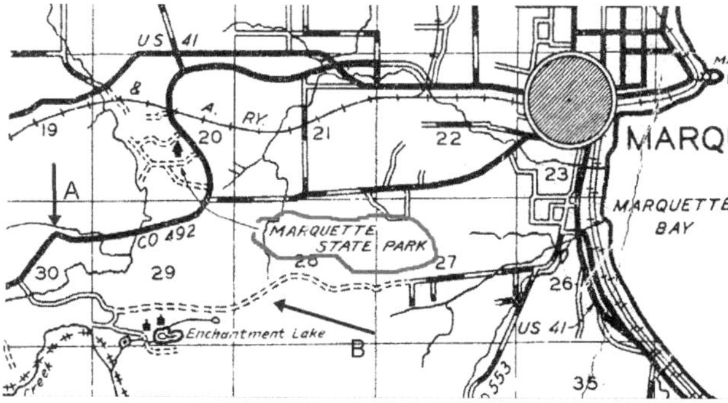

1941 Mich. Dept. of Conservation Map

Once the County Road 492 gets to the bottom of the S curve, it switches back to the Iron Mountain Rail Road grade for its route. If one looks out to the north side of the straight stretch when driving it, one sees the Plank Road grade. It is again on the north side of the the Rail Road grade After the present DSS&A RR grade was built on its present grade, the old grade must have been abandoned. It then became the new highway route to Marquette. As you see on the above map, before that time, the county road made a right turn just east of the Morgan Heights Sanitarium area and came into the Quarry area of south Marquette. Note that "B" shows the original County Road which probably was the road before it was built on the S curve.

Note on above map where the County Road went to the "B." route previous to the S curve route. If you look closely just left of the arrowhead in the photo, you will see pavement. It made a sweeping turn to the south going towards Marquette.

We now return to the straight piece of the Rail Road that has become the present 492 just west of the S curve. I am standing on 492 to take this photo of the Plank Road (double arrow) below. This was the George Rublein farm, and the original Wagon Road also came down north of here to the right of the photo coming south-west to cross at the bottom area of the S curve and going towards Marquette.

The photo on the cover is of the Plank Road as it comes to the bend where the Rail Road and the Plank leave the 492 straight stretch and go to north-west. Page 81 has the photo showing both grades leaving 492 going towards deadman's curve.

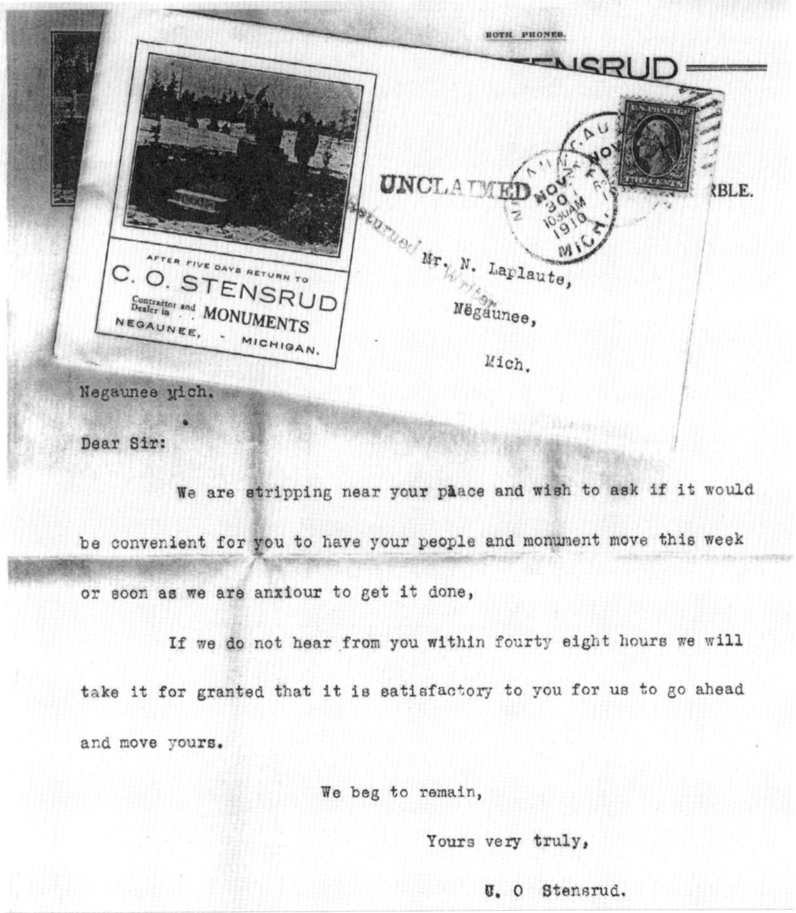

BOTH PHONES.

...NSRUD

...RBLE.

UNCLAIMED

NOV 30 10 30 AM 1910 MICH.

AFTER FIVE DAYS RETURN TO
C. O. STENSRUD
Contractor and Dealer in MONUMENTS
NEGAUNEE, - MICHIGAN.

Returned Writer

Mr. N. Laplaute,

Negaunee,

Mich.

Negaunee Mich.

Dear Sir:

We are stripping near your place and wish to ask if it would be convenient for you to have your people and monument move this week or soon as we are anxiour to get it done,

If we do not hear from you within fourty eight hours we will take it for granted that it is satisfactory to you for us to go ahead and move yours.

We beg to remain,

Yours very truly,

C. O Stensrud.

Here is a copy of the letter sent to those who had family remains in the old Negaunee Cemetery in 1910 between the Plank Road and the Rail Road. This is one of hundreds that had been returned and given to the Negaunee Historical Museum.
By 1913 a section of the Old Plank Road, The first Rail Road and most of the first cemetery were no more in the area of east Negaunee. The Cleveland Cliffs Iron Company gave the city $150,000 for the moving of the graves, according to the February 25, 1911, issue of the Ishpeming Iron Ore.

SECTION II: DEAD MAN'S CURVE & THE GORGE

The next section of the Railroad will take us from the west end of the straight stretch of Rublein's Farm, to "Dead Man's Curve" where the famous center line was put in by the County Road manager, K. I. Sawyer to prevent auto accidents. There was a sign at 492, but it has disappeared.

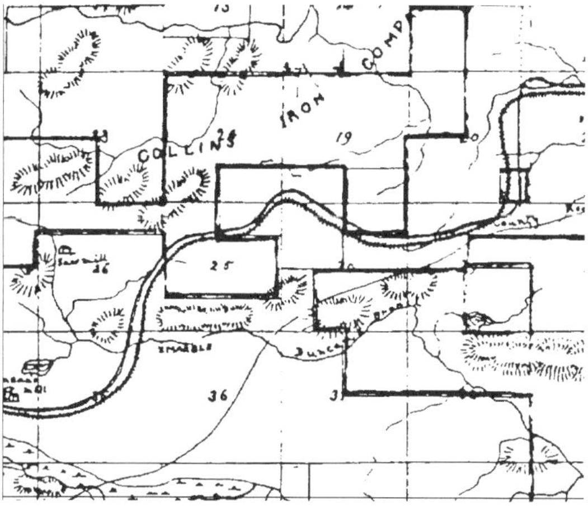

On this map, note a different County Road is shown than previously. It is not the Forge Wagon Road route. "County Road" is shown in upper right corner of the road at the bottom of the S Curve. It is interesting to note that the Wagon Road that went from the Cleveland Mountain to the Jackson Mountain, and then to the Jackson Forge and then on to Marquette is not on any of the early maps that we have printed that show the Plank Road or Rail Road routes. Yet, it was in use and carrying ore and blooms even as the Plank Road was being constructed.

When the Plank and Rail Road turn northerly, we don't have to go far to find "Dead Man's Curve." The next map shows where the newly built county highway joined the Plank Road going east towards Marquette. It joined the Plank perpendicularly, and made a sharp turn to the right onto the Plank, while going downhill on a non-banked curve.. You get the picture.

Credit for this next photo goes to Terraserver.com, for purchased custom map of the U. S. Geological Survey photo Lat. 46.53876 Lon: -87.45867.

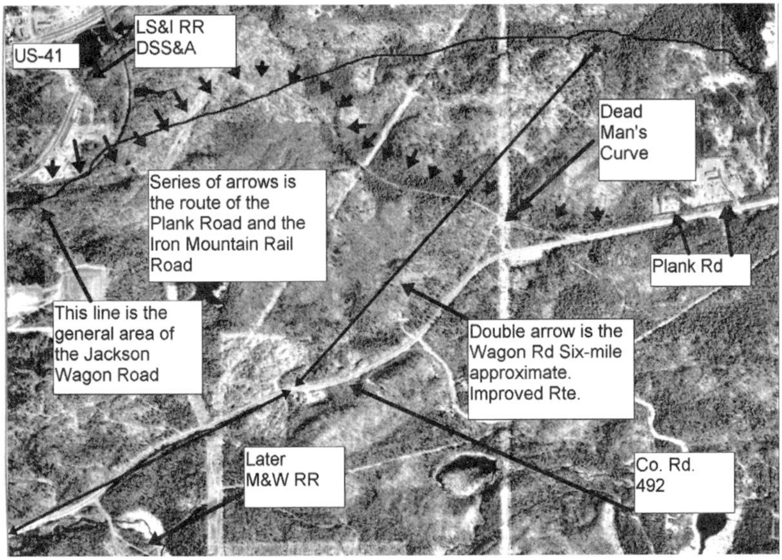

There was no Deadman's Curve on the Plank Road or First Railroad. It was created when it was decided to run the Morgan Heights auto road east and connect it to the Plank Road and Rail Road to go to Marquette. On July 8, 1916, the Iron Ore newspaper in Ishpeming noted that the new Marquette County highway to Marquette was very nice except for a very sharp curve about two miles east of Morgan Hill, which is just past the old Sanitarium. The old Morgan Heights Sanitarium is in the lower corner of the picture.

At Deadman's Curve it was very difficult to stay to the right and to keep from crossing into ongoing traffic and off of the highway to the left. There were many auto, and auto-wagon accidents, and even one motorcycle accident. Mr. Sawyer decided to do something new and that was to build up the far side of the road (east side) and bank it in the summer of 1917. This was a new thing for a highway. Not only that, he decided to paint a centerline on the road (which meant it was paved already in 1917.) It is believed to have been the first centerline in the nation.

One other item that I think should be mentioned that helped cause accidents is that many vehicles up to about this time were driving on the left as in England. The U. P. bought many Franklin autos as they were air-cooled and never had radiators to freeze up. Most of the Franklins had a steering wheel on the right side of the front seat and drove on the left, as did many wagons. It took a while to pass a law to drive on the right. I guess it meant wagons as well. If you look at the aerial map, you will see the present road now goes straight going east, and joins the old RR grade on the straight stretch. One wonders why it wasn't built that way at first.

Below is a photo of the County Road as it joins the Plank Road to form Deadman's Curve. The Plank Road is going right and left. The highway, going downhill, joined the Plank, then went to the right---and downhill. I don't see any banking today, at all. There is broken tar pavement, and no sign of a yellow line.

The RR grade is higher than the road in some of this area. Part of the old post and cable guard-rail on the left (drop-off) side of the Plank Road section going south-east is still visible.

There is a stream with old timbers between the "curve" and where 492 is today. It looks like the remains of the Rail Road bridge. It may be that the County Road through here may have been built on both the Plank Road and the Railroad for the three hundred feet or so out to the straight stretch.

We now turn our attention back to the Plank and Rail Road, going north-west from Deadman's Curve. The Plank Road and Rail Road are side by side, a little turn left, a little turn right, all the way up to a knob curve at the highest point. Vehicles are still driving on the Plank Road up to the knob area. There is a Rail Road Cut for a short distance. Then the presently traveled road goes off to the south at this point, towards the old Sanitarium. The Rail Road Grade is also plainly seen across a low spot. The Plank Road must be there also, but, being at ground level, it is grown over with trees and not visible. You can follow the Plank Road and Rail Road on the aerial map a few pages back.

After crossing the grade at the low spot, you can see the Lake Superior and Ishpeming Railroad in the distance, coming in from the north, and just a few hundred yards from where it crosses U.S-41. Now you will know where you are. The LS&I was only built about 1895, and it used the Plank Road space for a bit in this area. Industry in the area has obliterated any remains.

You will remember we left the new DSS&A grade on the big S curve. It has followed U.S-41 up the grade and crossed under the LS&I on the south side of U.S.-41. Going towards Marquette, it is on the right side of the highway, on the other side of a little pond and can easily be seen from your vehicle.

As we walk on the Iron Mountain Rail Road grade, we come to the LS&I tracks. We cross the LS&I tracks and get over to the DSS&A grade as both begin bending south towards the Gorge. Here is a present photo with the Original Plank and RR grade area on the right and the LS&I RR on the left. We are looking west. (The LS&I made its own cut into the gorge for its grade).

In this photo, the DSS&A grade is built over the Iron Mountain Rail Road grade. If you remember, the Plank Road was required to move north for the Rail Road by the "Award." The remains of that new Plank Road grade can still be seen, just north of the above photo, or to the right. The Plank Road grade is also much lower than the present RR grade. The next photo shows the Plank Road grade preserved very well for 151 years. We are facing toward Marquette in this photo, so the Rail Road is on the right.

Next, we will go south through the Gorge to see the 1848 Forge Wagon Road. Walk on the old DSS&A grade going south west from the gorge.(We hope this will become part of the very historical Heritage Trail.) You will walk next to a little lake on your right, probably part of the large Pine Hill Quarry. At the end of the lake you will find on the right, a road. It is not real big, but looks hand-built and has a raised area in the swamp area at the bottom, probably for a bridge. I have not been able to follow it from there, but there is a fine small cut from the swamp up to the DSS&A grade that would have been required to pull heavy loads on bare ground. It crosses where the DSS&A was later built, and appears to then go down into the swamp where it must have been crossed by both the Plank Road and the Iron Mountain Rail Road. We cover the swamp itself, to the left of the following photo, in the next chapter.

The Duluth, South Shore and Atlantic abandoned grade. The Forge Wagon Road is found at the end of the lake, to the right of the grade.

This is the top part of the Forge Wagon Road just down from the DSS&A grade.

This is part of the cut of the Wagon Road. The author hopes that someone will eventually walk this route and do a book on it. Further exploration going south-east should show that it continues towards the old Jackson Forge. Going towards Marquette will be difficult at first as it did not go through the Gorge, but through the area of the Pine Hill Quarry, where I think any historical remains have been destroyed. Where the Wagon Road crossed the Plank and Rail Road again at its northern-most point, the trail should be able to be found again.

Iron Mountain Railway and Iron Mountain Rail Road side by side leaving present 492 and going north west. This is part of Dead Man's Curve.

SECTION III: MORGAN FURNACE

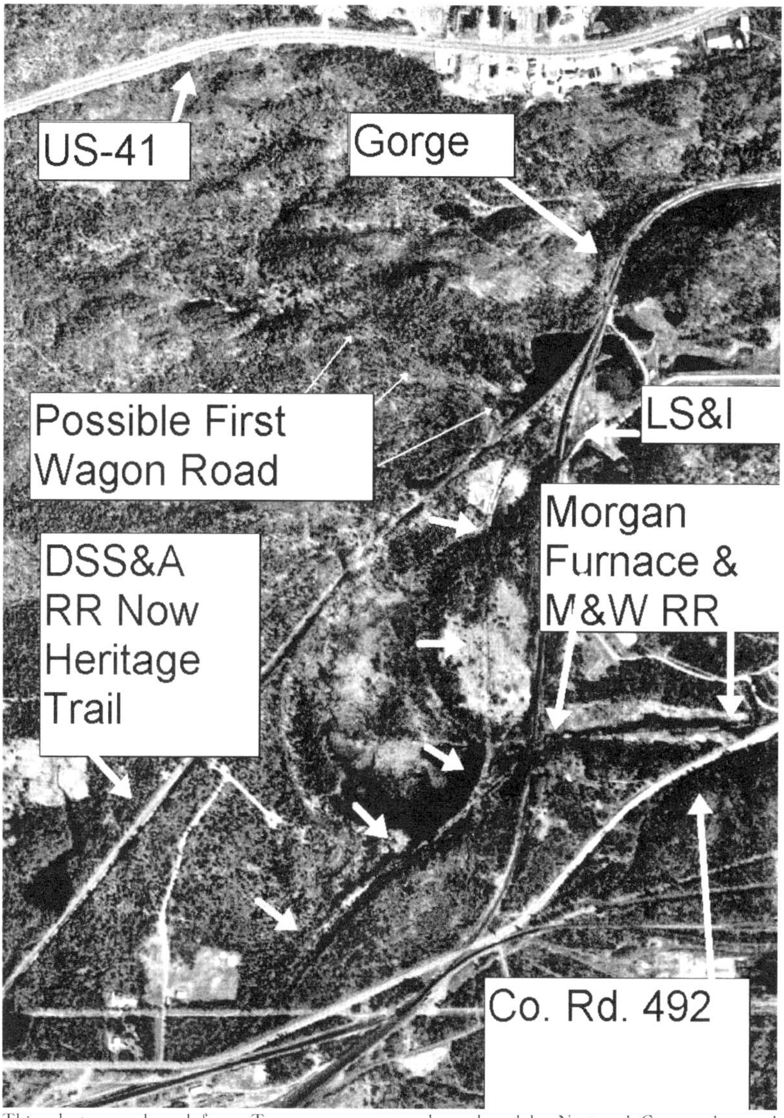

This photo purchased from Terraserver.com and produced by National Geographic and the USGS by Adventure Station.

The Gorge can be seen in the upper right corner of this Terraserver photo in the bend area. Walking south on the old

DSS&A grade (the future Heritage Trail), one walks past a small building at the south end of the gorge . It is the weighing station for every LS&I train coming and going to the docks in Marquette. At this point, the LS&I and the DSS&A grades separate. The Plank Road and the Iron Mountain Railroad enter a swamp about 20 feet below the current railroad grades, with hardly a mark left to see. On the aerial map, rather than cover up the grade with a pen, I have placed a series of arrows to show you where it is. This is the area where the locomotive kept sinking, and once came off the tracks as it tipped sideways. It is more swampy than usual these days. Those fine construction workers, the beavers, have taken over the area.

A strange phenomenon takes place. The Plank Road and the Rail Road, still running side by side, cannot be seen in the swamp from the DSS&A grade. I have taken several photos showing only trees, and a stream. From the air, however, the Railway and Rail Road are plainly visible as you can see. At the

end of the swamp, there is higher ground where I am standing for this photo. At this point, the Rail Road grade is higher than the beaver's pond, and can be seen coming south from the swamp. One can see why rock was brought in to stabilize the RR at this part of the grade. Upon reaching the higher ground the very flat Plank Road goes right up the gradual hill and the Iron Mountain Rail Road had to make a cut. It is not far from an ATV trail.

Staying on the Railway and Rail Road grades, going south towards Eagle Mills, the cut itself is rather lengthy, but not real deep, averaging about five to ten feet by my own estimate We again are looking north, back toward the swamp.

Clifford Stammer, standing in Iron Mountain Rail Road Cut south of bad swamp and Gorge. We are looking north.
The Iron Mountain Railway (Plank Road) is to the left.

The two Roads continued south and reach Morgan Creek. Both the Plank Road, and the Rail Road make a bend around the south east corner of Morgan Pond and head straight down to 492 in a south-west direction and come out just east of the Eagle Mills intersection.

Although Morgan Furnace was not built until 1863, it was served by a spur of the Iron Mountain Rail Road. The Furnace was just about 400 hundred feet east of it at the Morgan Pond. The furnace now lies almost under the LS&I overpass. The Marquette & Western RR came right by the Morgan Furnace starting in 1883, but it was too late for the Morgan Furnace as it burned down in 1881.

Here is a photo of the Furnace today. History is brought home when you think that this was built during the Civil War.

The LS&I trestle can be seen on the right, coming from Eagle Mills. The furnace is south-east of the trestle. Note the iron ore pellets under the trestle. Below is a close-up of the sign on 492 where the road comes in toward the furnace.

If you drive or walk in on the road from this sign you will see a memorial to Erin Taylor who disappeared and was found here shortly some days later. We print it here in her memory.

Memorial to Erin Taylor, found near the Morgan Furnace in August of 2000

Standing at the Morgan Furnace and looking north-west through the LS&I viaduct, we can see the edge of the Plank Road and the Rail Road while it makes its bend south west towards Eagle Mills from the north. Morgan Pond is just beyond them.

The next photo shows the straight grade from Morgan Furnace down to 492 at Eagle Mills. When walking, the old grade of the Plank Road with the Rail Road next to it can be seen below the present grade for part of the route.

Standing in front of Morgan Furnace, looking S-W toward Co. Rd. 492. Forground is the M&W Grade that joined the old Plank Road and IM RR grade just past the Birch Trees, on the right, coming south.

At 492, there are three separations of Rail Road grades. A power line goes through the area and has destroyed a part of the grade except for a ditch that can be seen from 492. But between the power line and 492, one branch goes off to the right, and must be the Plank Road. The middle one must be the Iron Mountain Rail Road. The one that goes almost straight is the M&W, the last Road to use that part of the grade. The M&W can be traced through the LS&I RR yard and goes west over M-35 just south of 492. It eventually went behind the present Negaunee Sewer Plant at 480, and to the mines.

An aerial photo of where the three Roads came out at 492.
(Purchased from terraserver.com. USGS & National Geographic.)

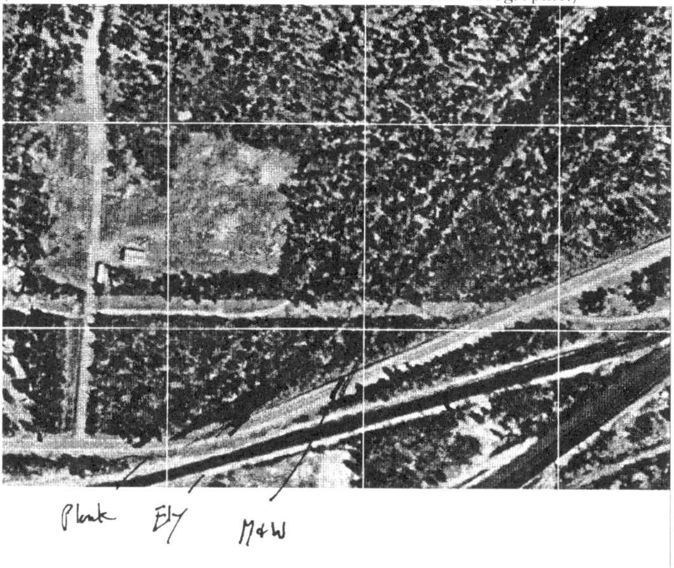

Next are three photos of where the Plank Road meets 492. We are looking north-east in the photo below.

Another view of the Plank Road going SW at 492. Rail Road grade is more southerly near top of this box.

Plank Road (Iron Mountain Railway)
Iron Mountain Rail Road

SECTION IV: ON TO NEGAUNEE

After the Iron Mountain Railway (The Plank Road) and the Iron Mountain Rail Road reached the 492 area, they continued west to Eagle Mills where Mr. Duncan's Large Saw Mill was located. There was a pond just north of it that was used to get water for his steam power. On the east end of Eagle Mills was a rise in the topography. There is none now, but the Referee Award Appendix will show the words "Summit of the Hill" on two occasions. It must be that the large sand bank on the west side of M-35 and north side of M-28 extended south originally and that both the Plank Road and Rail Road had to climb over it. The Referees Award gives the location of the summit as near station 448. The Rail Road crossed section 35 into 34 at station number 424.36. It would be just past Duncan's Sawmill, going towards Negaunee.

At the present Eagle Mills, the Plank and Rail Road then make their largest separation on the route, the Plank goes north-west toward the Jackson Forge and past where the new Negaunee Cemetery now is located The Rail Road went westerly after leaving Eagle Mills and then through some very swampy low land with the Carp River meandering through it. It was at this point that the grade, ties, and rails had to be carried and completed by manpower, the locomotive being too heavy in the building stage..

The Sharon Forge (Jackson Forge) was operating and they were the main builders of the Plank Road which was quickly being constructed. They wanted to ship ore efficiently through the Sault locks two years before the Rail Road was completed. That is no doubt why the Plank Road headed north in the cemetery direction (Remember,however, that the cemetery did not move here until 1912). However, as we know, the Sharon Co. sold out just before its completion, and new owners could not keep the Forge going. Thus, the wonderful, original Forge was never used by the Plank Road, and that also may be a reason why the Rail Road went west and never went to the Forge direction when constructed two years after the Plank.

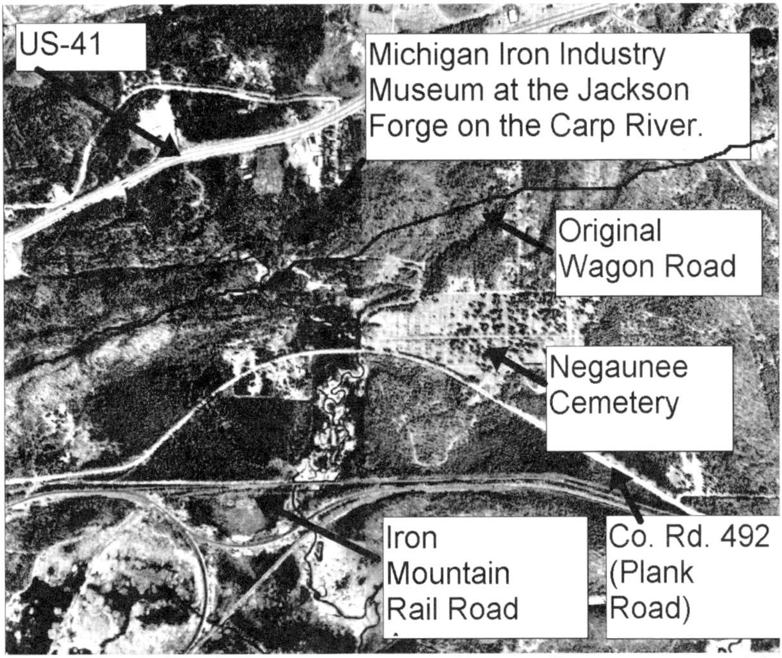

(Map purchased through Terraserver.com using Adventure Station.com supported by USGS and National Geographic.)

The map above shows the general location of the Jackson Forge. I have not been able to walk this portion of the Wagon Road shown on this map, but believe my hand drawn route is generally accurate.

Great details of the Jackson Forge are available at the Michigan Iron Industry Museum. There is no entrance charge and there is a fine store of books and other items.

The old Plank Road from the Eagle Mills area into Negaunee became the highway by the traveling public. Going toward Negaunee from the Forge area, we first find it crossing under the old LS&I RR viaduct. The LS&I didn't cross the Plank highway until about 1905, but it built a bridge over the then M-28 highway. It was also a "dead man's curve" where it crossed. The highway (old Plank) had to make a sharp jog left, and then, another, right. The viaduct is still there, but filled with sand, and can partly be seen. The present County Road 492 makes a turn to the right just before the viaduct and goes straight out to

U.S-41. You will see the old road going straight when 492 turns right. Walking in, you will see the old viaduct. Here is a partial photo of it. The Railroad is still active here, but the entrance into Negaunee changed in the 1970's.

If one takes the ATV trail from just a bit west of the old viaduct on 492, the LS&I tracks can be crossed and you will find the "old Plank" paved road continuing on west to Negaunee. Some of us wonder if the old planks are still preserved under the pavement. (See photo p. 38)

If you look to your left you will see the Rail Road grade. It is rather wide in places along the way, and may have been a place for ore car storage and preparation for the trip down to the Lake. It also has many cuts, some quite large and deep, the deepest seen since leaving Marquette's S curve. The Plank Road highway goes up a long hill. Here is a photo looking back east on the Rail Road Grade and showing part of a Snowmobile direction sign from a ways up the hill.

Upon reaching near the top of the Plank Road highway I could see the remains of the old cemetery. As there is a least one sunken grave, it seems that some bodies may have been left behind. In 1912, the last call went out for graves to be moved. The south ends of the cemeteries, and the Plank Road (Main Street), and Railroad are fenced off as this is the badly caved area. Caving had begun because the underground mines did not have a very good rock roof over their large mined-out cavities, that is, not good enough to support the surface.

There are still concrete enclosures of family grave-sites. Here is a partial one right next to the paved Main Street (old Plank).

The Plank Road, then named M-28, turned to the right at the top of the hill on to a street extension of Prince and Cherry Streets. However, the original Plank Road continued straight at the top of the hill and became Main Street We have several maps of the area, now all fenced in.

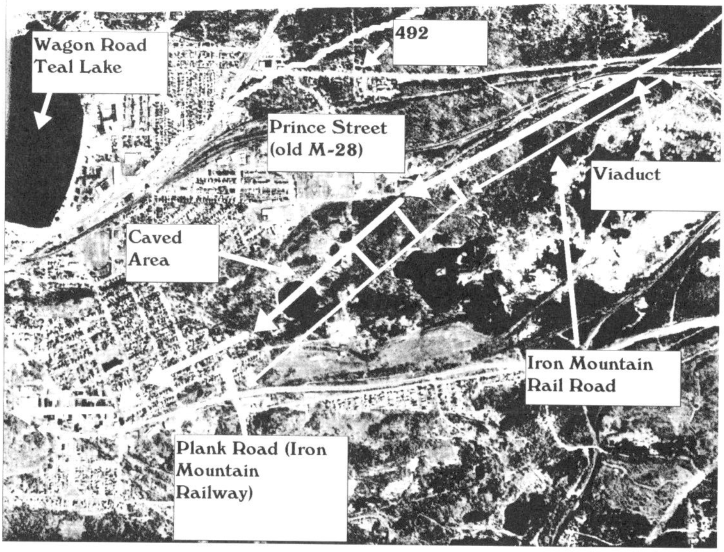

The thin arrows follow the route of the original Iron Mountain Rail Road, and the heavy arrows, the Plank Road Highway. The three lines perpendicular to the Plank and Rail Road represent where the original Negaunee Cemeteries were. The Rail Road, by then called the Duluth, South Shore and Atlantic, had to move east, and Prince Street was built west to tie into the Plank Road for a partial new route through the city of Negaunee. You can see on this aerial photo the large caved- in "perfect circle" area today where the Plank Road (Main Street) crossed directly over. The Railroad, although moved, because of caving ground had to be later moved a second time further east again. The wavy line in the left upper corner area on the photo above is the general area of the Wagon Road which operated from the Cleveland and Jackson Mines up to the time the Plank road opened in 1856.

This entire area from Prince/Cherry to the present corner of Main and Brown is fenced off from all public use. Here is a photo of where Main Street entered Negaunee from the old cemeteries area.

In the photo the Negaunee Historical Museum is the first large house seen on the left, at Brown and Main Streets. Here is also a close-up of the end of Main.

Here are several other maps showing the original Plank and Rail Road area. Note that the Proposed new Street is not called "Prince" yet. Note the many other streets here are now gone, including the "Old Baldwin Kiln Road" as the last street running North-West on the right center of the map. Note that the Present Baldwin Kiln is marked with "New". Also note that part of the Cemetery is shown. The area between "Old Kiln" and the cemetery and between the Plank and the Rail Road had many homes on it as well by 1890. We read in the May 5, 1887, Iron Herald that a new home on Main Street "near the cemetery being built for Waldman and Gripp has to be torn down and reconstructed" because of a bad storm.

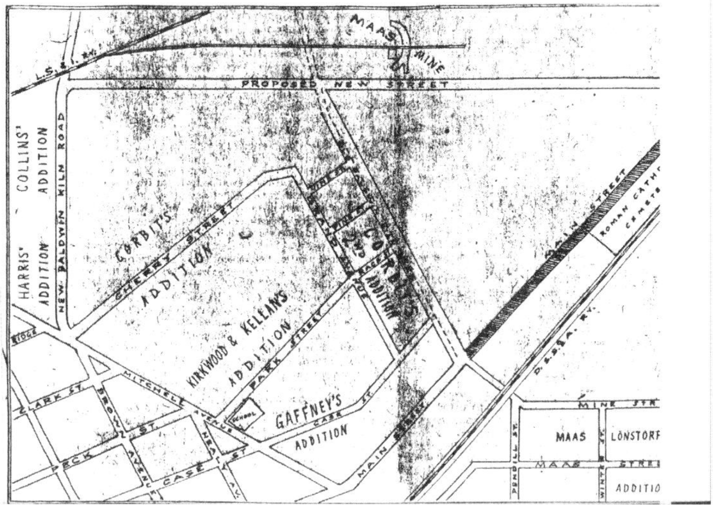

Note on this next map that Prince Street is now officially M-28.

On the preceding map is a better view of the cemetery land and the homes to it. I do not know what the large numbers are for. Note the caved in areas already. And most of all, note the many blocks of homes in the "Lonstorf and Mitchell Addition." This map seems to have been made at the peak of development.

The next map shows the "Additions" being abandoned and the years. This map seems to be from about 1960. Abandoned because of caving were:

Corbitt's Second Addition, and the Negaunee Race Course in 1928.
The Gaffney Addition shrunk gradually in 1941 and 1957.
The Kirkwood's and Kellan's Addition shrank in 1941, 1952, and 1960. Other records show the Lonstorf and Mitchell vacated in the 1920's. Also note that the DSS&A RR was abandoned. The homes were moved or torn down, evidently, as the caving spread out. Main Street must have been shortened several times.

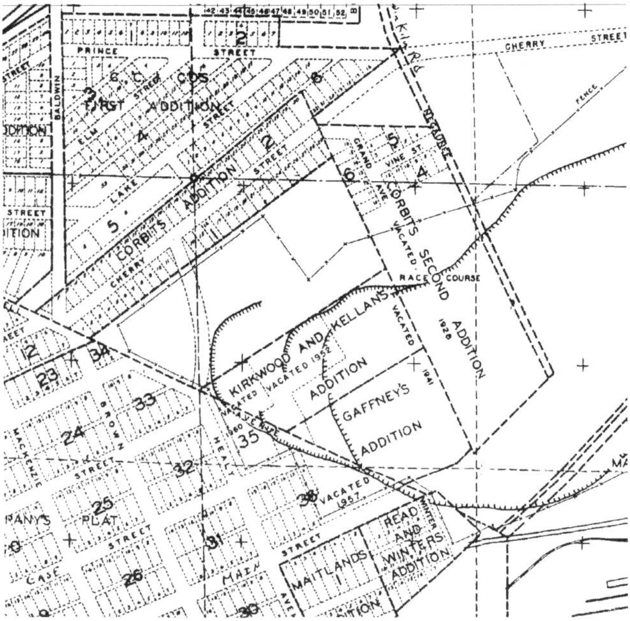

We now look at another aerial map of the area from the caving-in area to the Jackson Mine. The Iron Mountain Railroad originally probably ran down parallel to Main and tied in by the present Hockey Rink. After the original Rail Road became the DSS&A and it was moved east two times from the caving ground area, it used the area in front of Ann Street, south-east of Negaunee. The C&NW RR also used that route, joining from the south across today's 480 road. Both went where the present Hockey Rink is, then separated just before Silver Street.

The Pioneer Furnace was built during the Rail Road construction in 1855 and gave right-of-way for trackage. It had two stacks for making pig iron and produced an average of 450 tons (900,000 lbs) of pig iron each week, with much used for iron train wheels. It ran until 1893. Its location on the map below is at the bottom area, just to the right of the larger housing area.

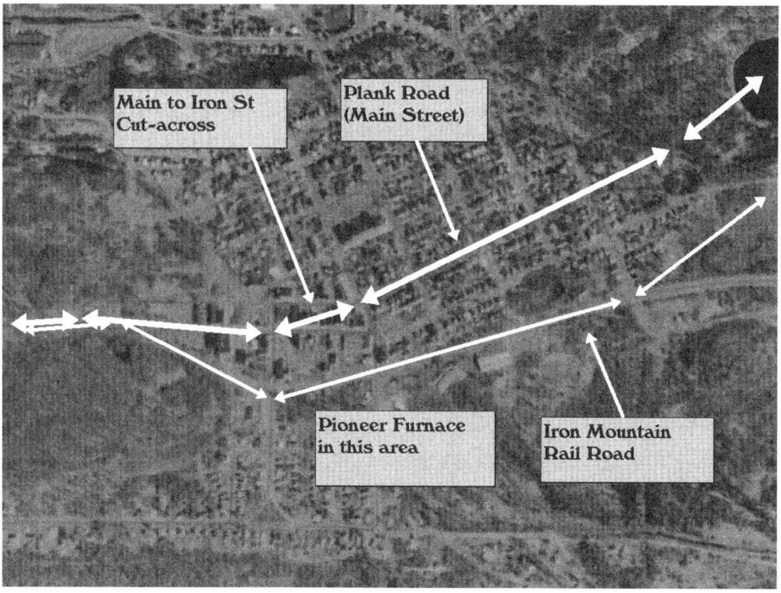

People have always been interested in how the Plank Road got from Main Street over to Iron Street and thence to the Jackson Mine. The shorter fat arrow, in town, just left of center is where the Plank Road went across the block diagonally from Main down to Iron Street. This probably accounts for there being no stores on Jackson Street because from Pioneer to Silver, the Railroad ran in the back of these businesses. The Range Bank, Lowensteins, and the Vista Theater that do extend to Jackson from Iron St. were not there at the time of the Civil War.

The reader will be interested in one very fine article concerning the building of the former Negaunee National Bank in 1924 when the bank was building the full block between Iron and Jackson. They found the remains of the Plank Road. The July 31st issue of the Daily Mining Journal reads: "Remnants of the historic plank road over which iron ore from the mines in this district was hauled to Marquette more than 70 years ago, were unearthed yesterday by workmen excavating beneath the foundations of the First National Bank building. Portions of the old corduroy upon which planks were spiked were found six feet below the present level of the street."

Note the abandoned blocks of homes north of the Pit # 1 area (The left side of the preceding map above the Rail Road arrows). This area has now been re-opened to the public after many years when it was believed it might cave in. Much was abandoned about 1970. I would like to call it a "Modern Ghost Town." Paved streets, concrete curbs, and sidewalks, as well as steps and concrete in people's yards are still very visible. Many have placed signs up indicating their family lived at certain addresses. It is recommended for visiting by the public. I hope you might also look around all the non-fenced in areas of the old Jackson Iron Mine, begun in 1848, thirteen years before the United States Civil War. Many forget how old our history is in the Upper Peninsula of Michigan. The property was first opened to the public in 2006, after purchase by the City of Negaunee from the Cleveland Cliffs Iron Company.

Credit for the next photo: Harlan Hatcher, A Century of Men and Iron, p. 142.

WOOD-BURNING LOCOMOTIVE DAYS, JACKSON MINE

Here is a photo of what looks like the Sebastopol Locomotive in every respect if compared with the sketch in Section I of this book. I have tried to identify the Jackson Mine area, but it may be a later mine as these engines operated for many years. The reader may check the maps for a similar curve. The smoke seems to be coming toward the photographer and if it is a north-west wind, then the engine would be coming a bit east at this point.

Another site at the Jackson Mine is where the Concentrator was. Here is a photo of it. It operated only from 1881 to 1886.

ACTIVE CONCENTRATOR at the JACKSON MINE, NEGAUNEE, OPERATED in 1880's

Its location is not fenced in and was located in the area just south of the high knob hill seen from US-41 to the south of Teal Lake and west of the old Mather "B" shaft and the present Negaunee High School. Some concrete remains are still there. It was seven stories high but was a failure and shipped only 13,000 total tons of concentrate. (Credit for photo and text: 13)

The Ishpeming and Negaunee Street Railway was also in the north Jackson Mine Area. After coming down Iron Street to Cyr Street where it turned north, it again went west on Mill Street. You cannot continue on Mill Street into caving ground area below the Mather "B" shaft. In that area, however, were the Street Car yards and the Negaunee and Ishpeming Street Railway and Lighting Company which provided the first electric power to the public starting in the 1890's. I have also walked most of the Street Car grade from the western side of the closed-off area but have not seen the following place mentioned in the Iron Herald on October 17, 1891:

"The Roadbed for the electric railway has not yet been entirely completed, the large cut near the concentrating works presenting considerable difficulty. There are monstrous boulders of all sizes to remove and the work is slow."

I realize these few pages of history did not relate directly to this book. However, they can be viewed on the maps, and to point them out without giving the reader any information about them would not be kind. At the present time, one cannot get to the concentrator from the Jackson Mine area as the Partridge Creek outlet has no bridge over it.

We now take a look at the very, very, old Jackson Mine. Here is a photo of what is believed to be Pit #1. This is where the Plank Road and the Rail Road should have entered the early mine. A larger pit with no water is behind it. In 2006, Pit #1 was opened to the public, but then closed because of required protection at the water's edge at the far side of the pit. The area is nice because it has a partial open drift for viewing. Hopefully, it will be reopened in the future.

Many other areas are open to the public. Do not enter any fenced-in areas. You may drive through the mine area from the Senior Center area in downtown Negaunee, or from M-28 as shown by the hand-drawn line starting in the bottom left corner of the map on the next page. You will probably be able to walk or bike, or skate-board and roller blade, a free paved Heritage Trail through the mine area. Signs will be posted for historic items.

On the aerial map below, the top right area arrows represent the Plank Road and the Iron Mountain Rail Road coming together on Iron Street. The very short arrow at that point shows where the both of them went into Pit #1. From Pit #1, the Rail Road then was built over the Plank Road Grade to the Cleveland Mountain in Ishpeming, and then went past the end of the Plank Road to the Lake Superior Mine at the "Little Mountain", or "Superior Mountain". Remember that the town names were made only after the Plank Road was already hauling ore to Lake Superior.

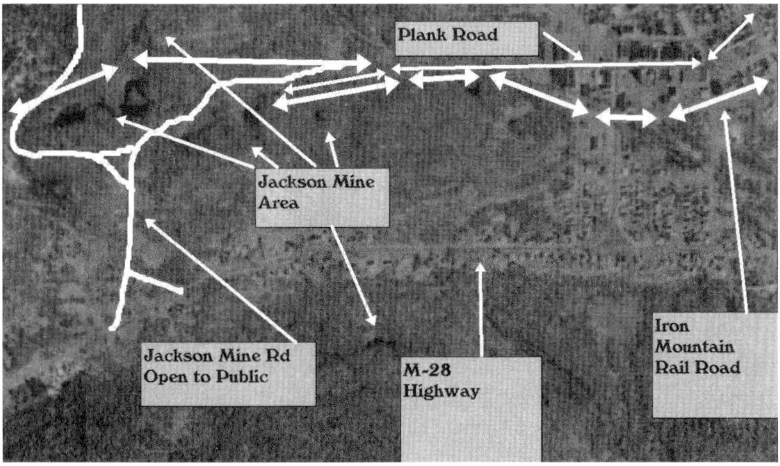

The Heritage Trail will historically follow the same route from the Jackson Mine to the Union Park area (as shown on the next map) on the Plank Road and Rail Road grade. As we follow the Railway and Rail Road though the Jackson Mine, we will walk on the beautiful new Heritage Trail. One unusual item along the trail are concrete boxes that have the old style electric porcelain insulators. We have no explanation. There was no electric railroad in this area.

SECTION V: TO TWO MINES IN ISHPEMING

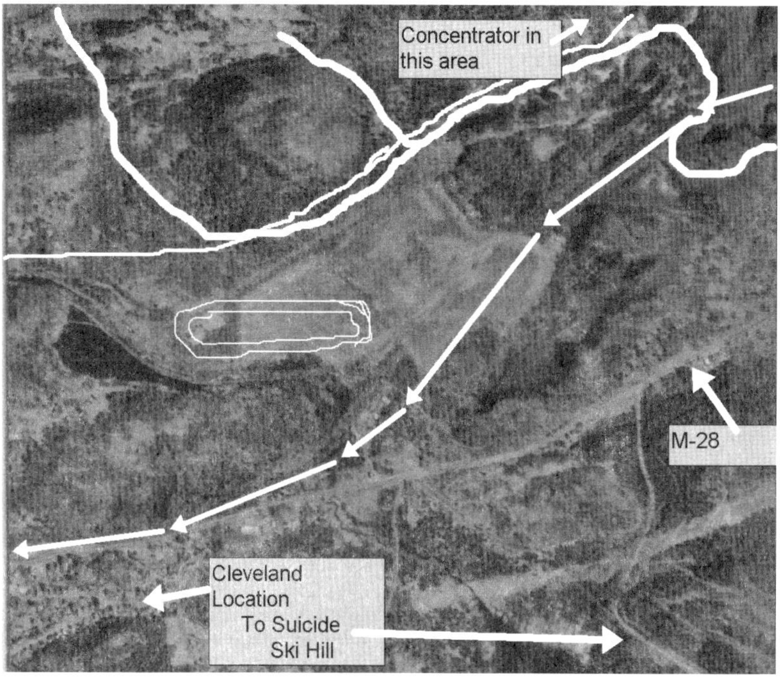

On the aerial map above we are leaving the Jackson Mine area. I have marked where the Concentrator was. The heavy drawn line is the old road to Negaunee and the Wawonowin Golf Course in Ishpeming. The thin hand-drawn Street Car line joins on the north side of the Golf Course road and parallels it for a long distance. We suggest you use a car or truck you don't wish to scratch up. At this time the paved roads are still quite grown in. Many walk. We are on the north-west side of the Jackson Mine.

The city of Ishpeming has not opened their CCI property to the public at this time. The city limit of both towns is a line almost straight down, but to the right of the middle of the previous photo. The drawn oval is the old favorite Union Park, used100 years ago for horse and bike racing, circuses, celebrations and much more. It is partially covered now by a former land-fill. The arrows once again are the route of the two Roads, but the Iron Mountain Rail Road is built over the Iron Mountain Railway (Plank Road) from Negaunee to Ishpeming.

After leaving the Jackson Mine area, the LS&I RR which has been traveling side by side with the Plank/Rail Road through the Jackson Mine area, separates at the former Howard Oil Company building, the larger building still standing. The LS&I went behind the building (ore cars seen), while the Plank/Rail Road went in front of it (left side of photo) going west.

(Photograph compliments of Arthur Howard)

Although the Heritage Trail has been on the Plank/Rail Road grade through the Jackson Mine, it crosses from that grade to the LS&I abandoned grade when the two railroads again come together on the way to Ishpeming just beyond the above buildings. The large building in the above photo can be seen in the aerial photo in the center of the picture just south of the landfill.

The Heritage Trail will be on the north side of the two grades, or the furthest from M-28. When this former LS&I grade crosses Davis Street in Ishpeming you will find the old round house on the right, with the concrete turnaround and areas to work under locomotives. The roundhouse for storing engines can also be seen by the concrete remains.

The old Plank Road and I. M. Rail Road grade follows along the LS&I grade, but veers off from the LS&I just before reaching Davis Street and goes west by south-west. An old trestle is missing across from the former Cleveland School. The grade goes between two homes on the north side of M-28, crosses it, and then enters a long railroad cut which widens as it goes along, holding several tracks at the far end. The Cleveland Mountain and former school is behind us, and to the left.

The Iron Mountain Rail Road built on the Plank Road grade from Neg. to Ish. and runs closer to M-28 than the later 1895 LS&I RR grade, shown with arrow. Looking east.

The next photo will be at the Cleveland Mine about a half mile to the left. The original Cleveland Mine was on the south side of Division Street, not on the north. You can see the early workings of the Cleveland behind the beautiful Iron Fence the Cleveland Cliffs Iron Company installed to protect the public from danger. The fence is on the edge of the road on the right side of the photo, as well as the early workings.

The hill to the right is also the hill with the "Bright and Shiny" Rock. If you are there at the top on a bright day, behind the

Cleveland School it is almost blinding. Just go up the hill on the road on which I'm standing (not the road on the right), and find the steps in the rock wall and the path up and past some more old mine workings. The street is Jasper.

This is the route of the original Plank Rd and over it, the Iron Mountain Rail Road. We are looking east from the original Cleveland Mountain Mine.

Below is a photo showing the location of the original "Cleveland Mountain" on the Bay De Noquette Railroad map of 1858.

In actuality, the richest part of the Cleveland was not across the tracks to the north. It ended up that the richest lode was on the south side of the Cleveland Mountain and on the right top section of Lake Angeline. Note also on the preceding map that the Cleveland Iron Mining Co. property is directly next to the Jackson Property on Section 1 & 2. On Section 3, Lake Lillie is now Lake Bancroft, and the larger lower lake on Section 10 is now Lake Angeline. According to Peter White, it was named for a beautiful young girl from Grand Rapids. (32, p. 257)

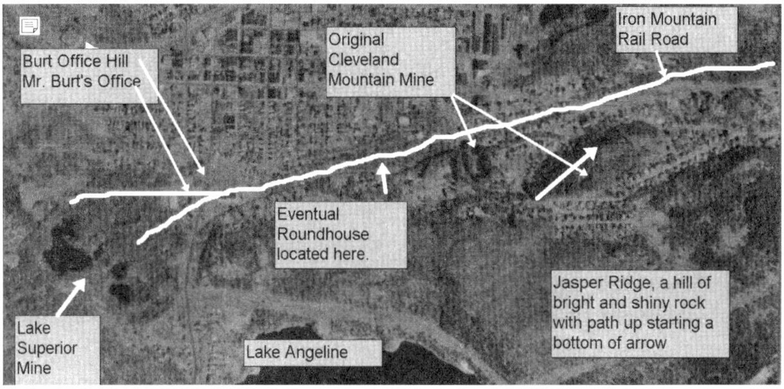

After leaving the Cleveland Mountain, the Railroad continued west on a new grade to the Lake Superior Mine. The Plank Road had stopped at the Cleveland Mine. The Lake Superior Mine began shipping only after the Iron Mountain Rail Road was completed in 1858. It is interesting to note that the Rail Road had two grades at the mine, to the left of the pit, and to the right.

Here is the photo from the Cleveland Mine heading towards the Lake Superior Mine.

The above photo shows one last remaining structure of the mine---a tall concrete pillar remains of some part of the mining requirements. I have yet to find it on a mine photo. This was the end of the Iron Mountain Railroad in 1858 on the north side of the pit, still three years before the Civil War broke out.

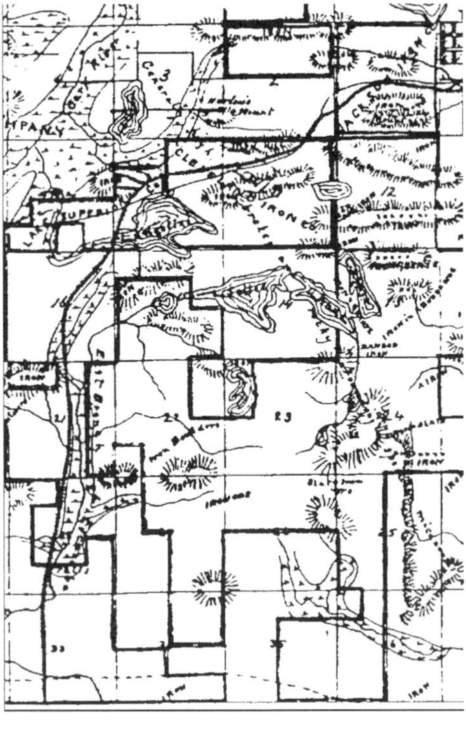

Eventually the Rail Road went to the National Mine area to several mines, and at the Winthrop Junction it headed west to L'anse and Houghton. On this map it is going to the National Mine area from the Lake Superior Mine.

Five Appendixes follow.

Appendix "A"
A Reprint of the "Referee Award"
Dated: October 2, 1855

Referee's Award.

In the matter of the difference between the Iron Mountain Ry., Co., and the Iron Mountain Rail Road Co., submitted to the decision of the undersigned by agreement dated the 13th of September 1855. The undersigned gives his answer and decision to be as follows, viz.,

That both the Railway and Rail Road are to remain as now located and partially constructed from their eastern terminii to Station 283. From thence to Station 386, the Railway to remain as at present located and built. From the last named Station to Station 396 the Railway is to be changed (according to agreement as follows, viz., at Station 388+82, rock to be cut northward a distance of twenty-seven (27) feet from present center of Railway to outside of cutting. At Station 390+30 the rock to be cut twenty-five (25) feet northward from present center of Railway to outside cutting and at Station 391 the rock to be cut northward a distance of ten (10) feet from present center of Railway to outside cutting and the Railway from Station 386 to 396 inclusive shall be moved northward to a new roadway as far as above specified cutting will admit and in such lines and curves as shall be found most expedient for said Railway when it is thus and there relocated and rebuilt from Station 396 to the Little Carp River. Near Station 432, the Railway is to remain as now located and built. From thence to the Summit of the Hill near Station 448 a new embankment and roadbed can be made for the Railway which can be moved northward not less than fifteen no more than one hundred feet at the point of greatest diverging from present line. From Station last named (Station 448) to the Jackson Iron Mountain the Railway to remain as at present located and built.

The Rail Road line is to be located as follows, to wit; From Station 283 to Station 286 it is to be to the southward of the line of the Railway as above described, and distant from it not less than twenty feet from center to center of the respective roads. At Station 287+50 the two Roads are to be not less than seventeen feet from center to center. From Station 289 to 356 the Rail Road is to be not less than twenty feet to the southward of the Railway. Measuring in all cases from center to center. From Station 356 to 396 the Rail Road is to lie to the southward of the Railway, as distant as practicable,, but at such points or places as may be necessary it shall have the right to be located by its Engineer at any distance not less than eleven feet from the Railway (from center to center). From last mentioned Station to Station 400 the Rail Road shall gradually diverge so that at last named point it shall be not less than twenty feet from the Railway and continue at that minimun distance until it reaches Station 432. From thence to Summit of the Hill near Station 448 the Rail Road is to lie at least fifteen feet to the southward of the Railway and twenty feet distant unless in the opinion of the Rail Road Engineer some important advantage to said Road is gained by a nearer approximation. From point last named to Station 450 the Rail Road is to diverge to southward till at last mentioned Station it is distant from the Railway at least twenty feet and continues so to Station 460. From thence to the Jackson Iron Mountain location the said Rail Road is to lie to the southward of the Railway at least fifty feet excepting at some point near said Mountain where in the opinion of the Engineer of the Rail Road it may be necessary to cross the line of the Railway and continue westward on the north side of the Railway, beyond which point of crossing this decision or reference does not extend westward.

In carrying out the specifications of this decision in building the respective route the following provisions and general rules are <u>to be strictly observed.</u> It is specially provided that a ditch is to be made and kept open between the two Roads from Station 386 to Station 393 of sufficient capacity to - the water freely at all seasons which may accumulate in that vicinity so that neither road may be injured by its overflow, etc. It is to be adopted as a general rule that in all places where the grade of the Rail Road is or may be above or below that of the Railway and the distance between tracks only eleven feet as herein before specified then one and one half feet in distance between tracks shall be added for each and every foot of such elevation. or depression. of said Rail Road and at the same rate for every fraction of a foot.

It is also a rule that in all places the grade surface or road bed of the Railway shall be left unbroken and unencumbered for a distance of four and one half (4½) feet outside of either rail. Excepting when said Railway may be changed as already provided for and specified herein, in which case, this rule shall apply after the road shall be so changed.

The Stations and distances mentioned throughout this decision are those used on the Railway route and survey and a map or plat here to appended marked "A" is referred to to show as near as may be the route and stations of the respective roads. And which may be definitely defined by reference to the descriptive survey and field notes which are minuted in the margin of the map or plat above mentioned. It is also a rule that the space between these two roads where such space is not more than twenty feet from center to center shall be held in common by both parties and roads and such space or ground shall not be used by either party or road for any purpose whatever if such use is objected to at anytime by the other party or road. The distances specified herein before as applying to space between roads is in every case to be construed as extending from center to center of each road between its rails.

This may certify that agreeable to the tenor of a certain agreement in writing made on the 13th of September 1855 at Marquette by and between the Iron Mountain Railway Company of the first part and the Iron Mountain Rail Road Company of the second part, etc., etc. I the undersigned as sole referee did on the 25th day of September make a personal survey of the present and proposed routes of the roads named in the agreement in company with the Engineer of each and that after a fair and full examination of the premises I have made a decision and award as contemplated in said agreement which is contained in the specifications provisions and rules of this document as above written.

Marquette, October 2, 1855.

Appendix "B"
The Survey of the Iron Mountain Rail Road Grade (Includes "A" mentioned in the Referee Award.)

Survey of the Iron Mountain Team and Railway.

Beginning at the stake on the west shore of Iron Bay on section 23
of township 48 North and Range 25 West and near the junction of
Washington and Lake Streets in the Village of Marquette on the
course bearing North 58° 47" W,
thence on a 1° 42' curve to the right 1164 feet to station 14.64
crossing the N&S qr. line of Section 23 at sta. 14.04 and 446 feet
N
of the centre of the sec., thence N. 77° 20' W. 3430 feet to sta.
48.44,
crossing the W. line of section 23 at sta. 40.88 and 1657 feet from
N. W.
corner of said section, thence on a 3° 59' curve to the left 556
feet to sta. 55.00
thence S. 80° 29' w. 500 feet to sta. 60.00 thence on a 3° 56'
curve to the right 266 feet to sta. 62.66.
thence N. 89° 06' w. 400 feet to station 60.00
thence on a 1°30' curve to the right 100 feet to sta. 67.00
thence on a 12° curve to the right 300 feet to sta. 70.00
thence on a 10° curve to the right 200 feet to sta. 72.00
thence on a 3° curve to the right 300 feet to sta. 73.00
thence on a 3° 03' curve to the left 800 feet to sta. 81.00
thence N. 82° 02' W. 1200 feet to sta. 102.

Crossing the W. line of section 22 at sta. 97-36 and 254 feet from the
N. W. corner, thence on a 2° 01' curve to the left 900 feet to sta. 111.00
thence on a 4° curve to the right 250 feet to sta. 113.50
thence on a 4° curve to the left 250 feet to sta. 116.00
thence on a 3° curve to the left 300 feet to sta. 119.00
thence on a 3° 36' to the right 250 feet to sta. 121.50
thence S. 79.49 W. 950 feet to sta. 131.00
thence on a 3° 59' curve to the right 600 feet to sta. 137.00
thence N. 76.17 W. 1900 feet to sta. 156.00 crossing the W. line of section 21
at sta. 150.06 and 345 feet from the N. W. corner.
thence on 3° 56' curve to the left 3500 feet to sta. 191.00
thence S. 36° 17' E. 1850 feet to sta. 210.50
thence on a 5° 01' curve to the right 2300 feet to sta. 233.55.
crossing the S. line of sec. 20 at sta. 223.18 and 741 feet from the N. W. S. E. corner, thence S. 79° W. 4066 feet to sta. 274.16
Crossing the W. line of section 29 at sta. 271.00 and 1237 feet from the N. W. corner,
thence on a 3° 58' curve to the right 933 feet to sta. 283.49
thence N. 64° W. 366 feet to sta. 287.15
thence on a 2° 55' curve to the left 587 feet to sta. 293.02
thence N 81° 10' W. 421 feet to sta. 297.23
thence on a 4° 02' curve to the right 1017 feet to sta. 307.40
thence N. 40° 03' W. 577 feet to sta. 313.17, crossing the North line section 30 at sta. 311.71 and 1490 feet from the N. W. corner
thence on a 4° 58' curve to the left 1912 feet to sta. 332.29
crossing the W. line of T. 48 N of R. 25 W sta. 328.67 and 343 feet from the S. W. corner of section 19

thence S 45° 06' W 497 feet to sta. 337.26 crossing the N. line of
section 25 at sta. 334.80 and 484 feet the N. E. corner
thence on a 3° 59' curve to right 780 feet to sta. 345.06
thence S. 76° 11' W 358 feet to sta. 348.64
thence on a 1° 01' curve to right 1000 feet to sta. 358.64
thence S. 86° 17' W 1136 feet to sta. 370.00
thence on a 3° 28' curve to the left 375 feet to sta. 373.75
thence S 73° 16' W 275 feet to sta. 376.50
thence on a 5° 30' curve to the left 650 feet to sta. 333.00
thence on an 8° curve to the left 400 feet west to sta. 337.00
thence S 5° 31' W 30 feet to sta. 387.30
thence on a 9° 18' curve to the right to sta. 390.30
thence S 33° 24' W 1444 feet to sta. 404.74
Corssing the W line of section 25 at sta. 391.31 and 3258 feet from
the S W corner, thence on a 3° 58' curve to the left 855 feet
to sta. 413.29
thence S 29' E 1467 feet to sta. 427.96
crossing the E line of section 35 at sta. 425.21 and 983 feet from
theN E corner, thence on a 5° 12' curve to the right1212 feet to
sta. 440.08, thence S 62° 37' W 364 feet to sta. 443.72
thence on a 3° 54' curve to the left 724 feet to sta. 450.96
thence S 34° 15' W 400 feet to sta. 454.96
thence on a 3° to the right 1834 feet to sta. 473.30
thence S 39° 21' W 3303 feet to sta. 606.33
Crossing the W line of section 35 at sta. 424.36 and 2850 feet
from the S W corner, thence on a 4° 01' curve to the right 781 feet
to sta. 513.54

thence N 62° 31' W 3690 feet to sta. 550.44. Crossing the W line of section 34 at sta. 541.03 and 1574 feet from the N W corner
thence on a 1°42' curve to the left 1306 feet to sta. 563.50
thence N 87° 02' W 250 feet to sta. 566
thence on a 2° 03' curve to the left 2200 feet to sta. 588.00
thence S 48° 01' W 1080 feet to sta. 598.80
thence on a 2° 59' curve to the right 1020 feet to sta. 609.00
Crossing the W line of section 33 at sta. 602.03 and 264 feet N of the qr. post,
thence S 78° 23' W 600 feet to sta. 615.00
thence on a 2° 01' curve to left 1000 feet to sta. 625.00
thence S 58° 10' W 2363 feet to sta. 648.63
thence on a 3° 05' curve to the left 442 feet to sta. 653.05
thence S 44° 34' W 2320 feet to sta. 676.25
Crossing the E line of section 31 at sta. 660.38 and 94 feet from the S E corner crossing the S line of T 48 N of R 26 W at sta. 662.05 and 94 feet from S E corner section 31
thence on a 3° 01' curve to the right 675 feet to sta. 683.00
thence S 65° W 2500 feet to sta 708.00
thence on a 3° 03' curve to the right 600 feet to sta. 714.00
thence S 73° 15' W 450 feet to sta. 718.50
thence on a 1° 58' curve to the right 450 feet to sta. 723.00
Crossing the W line of T 47 N of % 26 W at Sta. 722.78 and 2726

feet from N W corner to section 6, thence 87° 53' W 400 feet to sta. 727.00
thence on a 3° 03' curve to the left 300 feet to sta. 730
thence S 82° 59' W 1000 feet to sta. 740.00
thence on a 3° curve to left 300 feet to sta. 743.00
ending near the centre of section 1 in T 47 N and R 27 W
making 14.015 miles.

A. Brooks

Engineer

KNOW ALL MEN by these presents, that we the undersigned President and Directors of the Iron Mountain Railway Company, have adopted the above survey for the route of the train railway of said Company from the Village of Marquette, in the County of Marquette to the Jackson Iron Mountain in said County and have located the same according to said survey.
In witness whereof we have hereto affixed our hands and the seal of said Company this 28th day of April A. D. 1855.

C. I. Walker, President

/seal/

S. G. Caskey, Director
E. C. Walker, "
Alfred Russell, "

Appendix "C"
The Iron Mountain Rail Road Yard in Marquette

From the Lake Superior Journal, Marquette, September 13, 1856

"A day or two since, we were shown the plans of a new Machine Shop and Round-house for the new IM RR. The Machine and Blacksmith Shops are to be under one roof divided by a wall of stone. The whole building is to be 34 x100 feet and the walls of stone 20 feet high. Attached to the Machine Shop is to be an Engine House 18 x 26 feet, a stationary engine of 25 horse power will furnish the requisite power for the Machine Shop and other purposes. Within the walls of the Machine Shop will be the well or cisterns for the supply of water, also a portion of the tank for supplying the Locomotives, so arranged that hose can be attached to it and convey water to any part of the building in case of fire. The Car Shop which is at one end of the Machine Shop and in range of it, though several feet distant, is to be of wood, 34 x 74 feet and the same height as the Machine Shop. The Foundry, also of wood, will be 27 feet square and fitted up with the most modern improvements in this kind of building. The Round House is to be at the end of, and attached to the machine Shop, opposite the Car Shop, and will be built so as to accommodate four Locomotives, and arranged so that an addition can be made to it for 12 more. The Blacksmith Shop will be arranged for eight fires, with room for a trip hammer. The design of the buildings and their arrangement is as near perfection as anything of the kind we have ever seen and reflect great credit upon the designer, Mr. Cornelius Donkersly, formerly of Patterson, N. J. The Company has been extremely fortunate in securing the services of as accomplished a mechanic as Mr. Donkersly and we hope that they will not soon lose his valuable services. The Car Shop, we understand, will be put up this fall, and the stone for the Machine Shop, etc., will be quarried and cut during the winter, so as to have the buildings ready for use in the middle of next summer. The stone to be used is fine sandstone."

NOTE: The details of D, E, and F are largely copied from material found in the History of the Upper Peninsula of Michigan, Published by the Western History Company, in 1883. A reprint of it in 1972 by the Mid-Peninsula Library Federation in Iron Mountain, Michigan, should be found at most libraries.

Appendix "D"

The Early History of The Jackson Mine

The Jackson Mine is located on Section 1 Town 47, Range 27, within the corporate limits of the city of Negaunee, the whole of the section belonging to this company and comprising the original entry made under a permit from the Secretary of War in 1845. But little progress was made in the way of development, however, until the opening of the St. Mary's Falls Ship Canal, and the completion of the railroad to the "Jackson Mountain," as the locality was then called, some ten years afterward.

In one particular, this celebrated mine outranks all the others in the district; it was the first discovered iron ore deposit in the region, and the company is the oldest corporation and the mine was the earliest worked, and from the ore of the Jackson the first iron was made ever obtained from the deposits of Lake Superior.

The company derives its name from the city where in 1845 the association was originally formed. The parties to this organization were Abram V. Berry, P. M. Everett, S. W. H. Munroe, A.W. Ernst, and F. Farrand, of Jackson, Michigan. In 1848 the name assumed was the Jackson Mining Company of Michigan. A few years later it became the Jackson Iron Company. A short time previously, however, a forge had been constructed on the Carp River, in which the first iron ore was made into blooms. The daily product of this forge when in full blast is about three tons of iron. The manufacture of blooms in this forge was abandoned in 1856. Until 1864, the Jackson,

Cleveland, and Lake Superior were the only companies which shipped ore from the region.

Geologically, the formation at the Jackson possesses great irregularity, and the occurrence of the ore deposits is largely a matter of conjecture, and their continuance when formed is only determined by the working. Extensive explorations are constantly made in the mine by means of trial shafts, crosscuts, and drifts made here and there, seemingly to the novice, at random, but really located with the greatest care by Capt. Merry, whose twenty years experience in charge of the mines and company's interests have given him a degree of familiarity with all the difficulties to be encountered, and a knowledge of the peculiarities of the formation.

In the east end mine, No. 7 and old No. 1 pits, the product is the best grade of hematite, possessing a percentage of manganese and also of chromium. Even here occur veins or pockets of hard ore, alternating with the soft ore deposits. The formation is exceedingly contorted and broken up, the veins doubling and folding in a manner nearly defying definite determination. The somewhat friable, disintegrated character of the rock has added an element of perplexity to the problem of how best to find and win the ore.

The South Jackson is the most easterly pit of the Negaunee hematite openings and is in the southeast part of the company's property; the section corner post stands upon the edge of the south pit and the line west from it crosses the opening, cutting off the north part of the Iron Cliff, and the line from the State south cuts off a portion of the pit to the east, belonging to the McComber Company. The openings of the pits at the South Jackson extend about a quarter of a mile in length and are about sixty feet in width. The hoisting is done from three pits by a separate engine for each pit.

The ore is carefully picked, and although somewhat mixed as it comes from the mine, the rock is all sorted out. The drilling is all done by hand; the company has no compressor and uses no power drills. Capt. Merry does not regard their use as likely to prove economical....neither is the diamond drill used to any great extent. The difficulties arise mainly from the peculiar nature of the rock; the small loose pieces of hard jasper wear out the diamond so rapidly that it is almost impossible to proceed. In its lifetime, closing in 1924, it shipped 4,362,629. tons. (8)

Appendix "E"
The History of The Cleveland Mine

The Cleveland Mine comprised 2,200 acres by 1880. The workings are mainly in the north part of the east half of Section 10 and the west half of Section 11, adjoining the Lake Superior and NewYork Mines. From the History of Upper Peninsula of Micigan, published in 1883, we read:

"These lands were first taken by a 'permit' from the War Department by Dr. Cassels, of Cleveland, in 1846. This gentleman visited Lake Superior in the interest of the Dead River Silver and Copper Mining Company of Cleveland. Dr. Cassels was directed to the locality by Mr. Everett and party, who had previously secured the Jackson Company's lands, lying adjacent to the east. Other parties subsequently entered upon the lands and instituted a pre-emption or mining right claim, but the title was finally accorded to the Cleveland Company."

A. P. Swineford (29) tells us in 1869 that the officers of the company were: George Worthington, President; John Outhwaite, Vice President; S. L. Mather, Secretary and Treasurer; Jay C. Morst, Agent, and F. P. Mills, Mining Superintendent.

Mr. Swineford tells us that in 1869 that ore is shipped about half and half on both the MH&O (successor to the Iron Mountain Rail Road) and the Chicago and Northwestern Railroad. The ore being mined at this time comes now from the following places on the south and north side of Division Street:

No. 5 Opening. "This no. 5 is a very large mine, from which can be raised an incalculable amount of the very best granular ore...." This vein goes west from the no. 5 along the base of "Little Mountain." (This must then be the ore body that the Lake Superior Mine is obtaining as it has always been suggested that they are at the "Little Mountain" to the west of the Cleveland.

No. 4 Bottom. We learn that the shaft is down 50 feet and there is an ore body at the bottom of twenty feet. Ore is raised by steam in a "skip" which is emptied into a small car on a trestle and discharged into railroad cars. It is planned to make a tunnel of 200 feet to No. 1 as it is believed ore is in that direction.

No. 1 Cut. This is an open cut, 250 feet long, and 20 to 50 feet in width. The bottom is solid ore between two hanging walls of rock. The ore is slate of about 65% iron.

West Vein of No. 1. This shaft is separated from No. 1 by a twenty-foot rock wall of rock, and it is believed that the bottom will also be solid ore. (One of the pits seems to have been filled up in order for the CCI Mining Company to build its steam power plant on the north side of Division Street in the 1950's.)

No. 2 and 3 pits are just west of No. 1 and connected by a narrow vein of iron ore. They are now one opening and are being worked on the same levels, of which three are active. Two drifts are being driven from two of the levels in a southwesterly direction. There is a 75-foot shaft at the bottom of the pit, and a shaft house and hoisting machinery. The ore is about 450 feet in length and from 50 to 150 feet wide. Ore is raised in two skips that each hold 3.5 tons and they are raised and lowered alternately by power from the Engine House with two pairs of double engines of 80 horesepower each. One pair hoists and pumps out No's 2 & 3, and the other pair is used to hoist cars up from the "Incline Mine." The building is 50 x 76 feet, with a machine shop in the upper story.

The "Incline Mine." Cars are raised and lowered into this 65 foot-long shaft which is 45 feet below the surface. The shaft will go 35 more feet deeper and a 500-foot drift is being built from the bottom to No. 1, 2, and 4. Ore found will be raised from the incline shaft.

The mine has built a large store, 26 x 80 feet with sales of almost $100,000 a year from the location. A number of "substantial frame dwellings for the employees of the company" are replacing "the miserable log cabins originally thrown together for the accommodations of the miners and their families." Mr. Swineford goes on to tell us that in 1869 there are about 250 men employed, most of whom work on "contract" and get paid by the amount of ore produced.

Statistics from 1854, the first year of production, through 1868 shows that total tonnage shipped was 285,215 tons.

In 1868 the shipping via Marquette was 33,155 tons, and from Escanaba (C&NW) there were shipped, 64,715 tons of ore. Sales to local furnaces were 4,343 tons. (It is taken for granted that all of the tonnage went to furnaces, but the balance of ore went to their own Pioneer Furnace in Negaunee.) I am unable to compare a shipping figure for the Cleveland Mine by 1924 with the Jackson and the Lake Superior as my Crowell and Murray book has somehow omitted the mine entirely.

Appendix "F"
The Lake Superior Mine

Although The Lake Superior Mine was organized in 1853, it waited for the Iron Mountain Rail Road to be completed to the mine in late 1857 and in that same year, the mine began operations as a mine, and shipped ore beginning in 1858. John Burt was not associated with the Plank Road at all, but in 1851 was already proposing the Rail Road with Heman Ely. (10) The company began with capital from stock sold of $300,000. Eventually $500,000 of stock was sold. The size of the property was only 120 acres originally in Sections 9 & 10 of Town 47,

Range 27. Eventually the mine owned 6,551 acres by 1883, but often in different areas. (14)

According to early history, Mr. Burt's Mine was at the "Little Mountain." The first mining was in open pits. At the Lake Superior, more ore was on the surface in a larger area, and not "so irregular and complicated as the Jackson." (Mining Journal, December, 1880) The visible ore was in a large body in a vertical "U" shape with other rock in the middle, but with the two top edges showing. About 400 to 500 feet was between the two exposed parts. Much diamond drilling was done, and after the open pits were worked down to about 105 feet, horizontal drift and rooms of ore taken out at the edges of the pit at the 105 foot level, called the first level. Three shafts were then sunk to go below the first level, of which one was at an angle rather than vertical. Already, in 1880 there were underground chambers over 400 feet long and 50 feet wide with ore left behind to form a roof over the drifts below. Note was made that ore is taken from a drift, often above a drift and falls down into the drift and into an ore car, is taken to the shaft, and raised to the surface in a bucket called a "skip." By 1880 the mine was down to the 280-foot level. The author of the article also notes that there were drifts in every direction, and that once all the ore possible could be taken, the ore between the drifts was removed, starting at the top level, and allowing the surface to cave in the mined-out sections.

Eventually a body of hematite ore was found at the Lake Superior Mine, and a large, two skip inclined shaft was put down into it on the south side of the original mine. The Hematite eventually went under Lake Angeline, which for many years was drained completely for the purpose of mining the ore.

The Mine was mostly on the west side of South Pine Street extending over to the Barnum Mine property. However, there was an "A" shaft on the east side of Pine, in the area of the retiree Pioneer Bluff Apartments. It went down and found ore and mined quite a bit, but had to leave much ore as a roof which was covered by sand or gravel only. To get this ore left behind, it was decided to make an open pit here and then to take the roof of ore left behind. The mine was in the process of infringing under South Pine Street with the street getting narrower and narrower. The mine then made a timbered roof over the ore body roof to allow both mining and rebuilding the street. The pump at "A" shaft also pumped Lake Angeline water to the city of Ishpeming's water system. The steam was piped to the engine from the boilers at the number 2 shaft.

The reader may be interested in an article written by a reporter from the Marquette Plain Dealer and published on October 1, 1868. The author notes that this mine is "the biggest thing in iron in this region." The reporter continues: "We donned a miner's suit, and provided with a lighted candle descended the main shaft by ladder. An intelligent miner guided us down and enlightenend us on the principles of the pumping, hoisting, and other mining machinery, we found in the shaft. We were surprised by the amount of work that has been done here. Drifts run in every possible direction. The floors are covered with three inch plank, there is a cistern 16 feet in depth and with the pumping machinery combined to keep the drifts free of water." The reporter also notes that they found two horses that live underground, busy working underground, and that the mine is now 100 feet deep, and there are the three shafts mentioned above. (And the mine is only 11 years old at this time.)

We should mention that the Lake Superior Mine in 1882 shipped the largest amount of ore of any one mine in America. (14) The tonnage was 296,509. for the one year.

By 1924 the Lake Superior Mine had taken from the earth, 18,875,408 tons of iron ore. (8)

Above: Where the Plank Road cut down from the end of Main to Iron Street on the back of the Iron Street lots.

This is a view of the Heritage Trail from the Jackson Mine going toward Ishpeming. Note one of many concrete electric boxes on the right. Its use is unknown and it is not the area of the electric Street Railway. We are on the original Plank Road and First Rail Road grade.